Vascular Biology of the Placenta

Integrated Systems Physiology: from Molecules to Function to Disease

Editors
D. Neil Granger, *Louisiana State University Health Sciences Center*
Joey P. Granger, *University of Mississippi Medical Center*

Physiology is a scientific discipline devoted to understanding the functions of the body. It addresses function at multiple levels, including molecular, cellular, organ, and system. An appreciation of the processes that occur at each level is necessary to understand function in health and the dysfunction associated with disease. Homeostasis and integration are fundamental principles of physiology that account for the relative constancy of organ processes and bodily function even in the face of substantial environmental changes. This constancy results from integrative, cooperative interactions of chemical and electrical signaling processes within and between cells, organs and systems. This eBook series on the broad field of physiology covers the major organ systems from an integrative perspective that addresses the molecular and cellular processes that contribute to homeostasis. Material on pathophysiology is also included throughout the eBooks. The state-of the-art treatises were produced by leading experts in the field of physiology. Each eBook includes stand-alone information and is intended to be of value to students, scientists, and clinicians in the biomedical sciences. Since physiological concepts are an ever-changing work-in-progress, each contributor will have the opportunity to make periodic updates of the covered material.

Published titles

(for future titles please see the website, www.morganclaypool.com/page/lifesci)

Vascular Biology of the Placenta
Yuping Wang and Shuang Zhao
www.morganclaypool.com

ISBN: 9781615040452 paperback

ISBN: 9781615040469 ebook

DOI: 10.4199/C00016ED1V01Y201008ISP009

A Publication in the Morgan & Claypool Life Sciences series

INTEGRATED SYSTEMS PHYSIOLOGY: FROM MOLECULES TO FUNCTION TO DISEASE

Book #9

Series Editor: D. Neil Granger and Joey Granger, Louisiana State University

Series ISSN

ISSN 2154-560X print

ISSN 2154-5626 electronic

Vascular Biology of the Placenta

Yuping Wang
Shuang Zhao
Louisiana State University

INTEGRATED SYSTEMS PHYSIOLOGY—FROM MOLECULES TO FUNCTION TO DISEASE #9

MORGAN & CLAYPOOL LIFE SCIENCES

ABSTRACT

The placenta is an organ that connects the developing fetus to the uterine wall, thereby allowing nutrient uptake, waste elimination, and gas exchange via the mother's blood supply. Proper vascular development in the placenta is fundamental to ensuring a healthy fetus and successful pregnancy. This book provides an up-to-date summary and synthesis of knowledge regarding placental vascular biology and discusses the relevance of this vascular bed to the functions of the human placenta.

KEYWORDS

placenta circulation, trophoblast, angiogenesis, vasoactivity, placental stem cells

Contents

CHAPTER 1

Introduction

The placenta is literally the "tree of life." The derivation of the word placenta comes from Latin for cake (*placenta*), from Greek for flat, slab-like (*plakóenta/plakoúnta*); and from German for mother cake (*mutterkuchen*) all referring to the round, flat appearance of the human placenta. Structurally, the placenta is a hemochorial villous organ. Functionally, the placenta is a highly complex machine: (1) it acts like a lung in the exchange of oxygen and CO_2; (2) it works as a digestive system, absorbing all necessary nutrients for fetal development and growth; (3) it functions as a kidney to remove wastes; and (4) it behaves as an immune barrier that protects the growing fetus from antigen attack from the maternal system. The placenta is also an important endocrine organ producing many hormones and growth factors that regulate the course of pregnancy, support and promote fetal growth, and initiate parturition. However, all these tasks depend on normal vascular development within the placenta itself. Normal placental vascular development ensures a healthy pregnancy outcome, whereas insufficient or abnormal placental vascular development will compromise pregnancy outcomes both of the mother and the fetus, with complications that include preeclampsia and intrauterine fetal growth restriction. The functional unit of the placenta is the chorionic villus, which contains the layers of syncytiotrophoblasts/cytotrophoblasts, villous stromal, and fetal vascular endothelium that separate maternal blood from the fetal circulation. We have collected and updated available information on current understanding of the vascular function in the human placenta.

• • • •

CHAPTER 2

Placental Blood Circulation

The placenta is a unique vascular organ that receives blood supplies from both the maternal and the fetal systems and thus has two separate circulatory systems for blood: (1) the maternal-placental (uteroplacental) blood circulation, and (2) the fetal-placental (fetoplacental) blood circulation. The uteroplacental circulation starts with the maternal blood flow into the intervillous space through decidual spiral arteries. Exchange of oxygen and nutrients take place as the maternal blood flows around terminal villi in the intervillous space. The in-flowing maternal arterial blood pushes deoxygenated blood into the endometrial and then uterine veins back to the maternal circulation. The fetal-placental circulation allows the umbilical arteries to carry deoxygenated and nutrient-depleted fetal blood from the fetus to the villous core fetal vessels. After the exchange of oxygen and nutrients, the umbilical vein carries fresh oxygenated and nutrient-rich blood circulating back to the fetal systemic circulation. At term, maternal blood flow to the placenta is approximately 600–700 ml/minute. It is estimated that the surface area of syncytiotrophoblasts is approximately 12m^2 [1] and the length of fetal capillaries of a fully developed placenta is approximately 320 kilometers at term [2,3]. The functional unit of maternal-fetal exchange of oxygen and nutrients occur in the terminal villi. No intermingling of maternal and fetal blood occurs in the placenta. Figure 2.1 illustrates (1) the relationship of the uterus, placenta, and the fetus, and (2) the directions of blood flow from mother to the placenta as well as fetal blood flow from the placenta to the fetus.

2.1 MATERNAL-PLACENTAL BLOOD CIRCULATION

Uteroplacental circulation is not fully established until the end of the first trimester. Although the exact mechanism of how the uteroplacental circulation is established is not completely understood, two theories have been proposed. The first theory is that in the first trimester, endovascular trophoblasts migrate along the decidual spiral arteries, invade the vessel walls, and create a path for maternal blood to perfuse the placenta intervillous space. This theory is supported by the presence of endovascular trophoblasts in the decidual spiral arteries of the first trimester placenta [4,5]. The second theory proposes that trophoblasts invade decidual spiral arteries and form trophoblastic plugs. These trophoblastic plugs obstruct maternal blood flow into the intervillous space and prevent flow until the end of first trimester of pregnancy (10–12 weeks). The plugs then loosen and permit

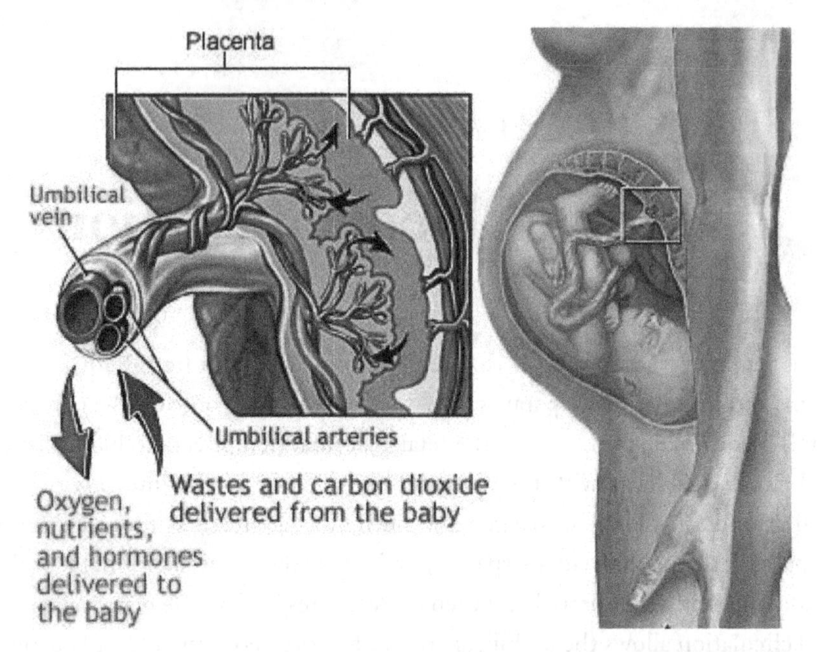

FIGURE 2.1: An illustration of human pregnancy and directions of the placenta blood flow. The right panel shows the relationship of the uterus, placenta, and fetus during pregnancy. The left panel shows the directions of blood flow from mother to the placenta and fetal blood flow from the placenta to fetus. The placenta provides the fetus with oxygen and nutrients and takes away waste such as carbon dioxide via the umbilical cord. (The figure is adapted from American Accreditation HealthCare Commission (A.D.A.M., Inc.) website: (www.urac.org). Used with permission of A.D.A.M.

continuous maternal blood flow into the intervillous space. This theory is based on the observations of *ex vivo* histologic analysis of hysterectomy specimens of first-trimester placentas, in which plugs of trophoblasts were found either partially or fully obstructing or filling the vessel lumen of decidual spiral arteries [6]. Although the two theories diverge as to whether or not invading trophoblasts 'plug' the arteries to prevent blood flow into the intervillous space, it is clear that the genesis of uteroplacental (maternal-placental) blood flow during the first trimester is a dynamic and progressive process.

Normal early placental development results in transformation of spiral arteries that extend from the decidua (the layer of tissue lining the uterus) to the muscle layer. Most textbooks provide the classic description of the placenta circulation based on studies of second-, or third-trimester placentas. As shown in Figure 2.2, maternal blood enters the placenta through the basal plate endometrial arteries (spiral arteries), perfuses intervillous spaces, and flows around the villi where exchange of oxygen and nutrients occurs with fetal blood. It has been estimated that there are about

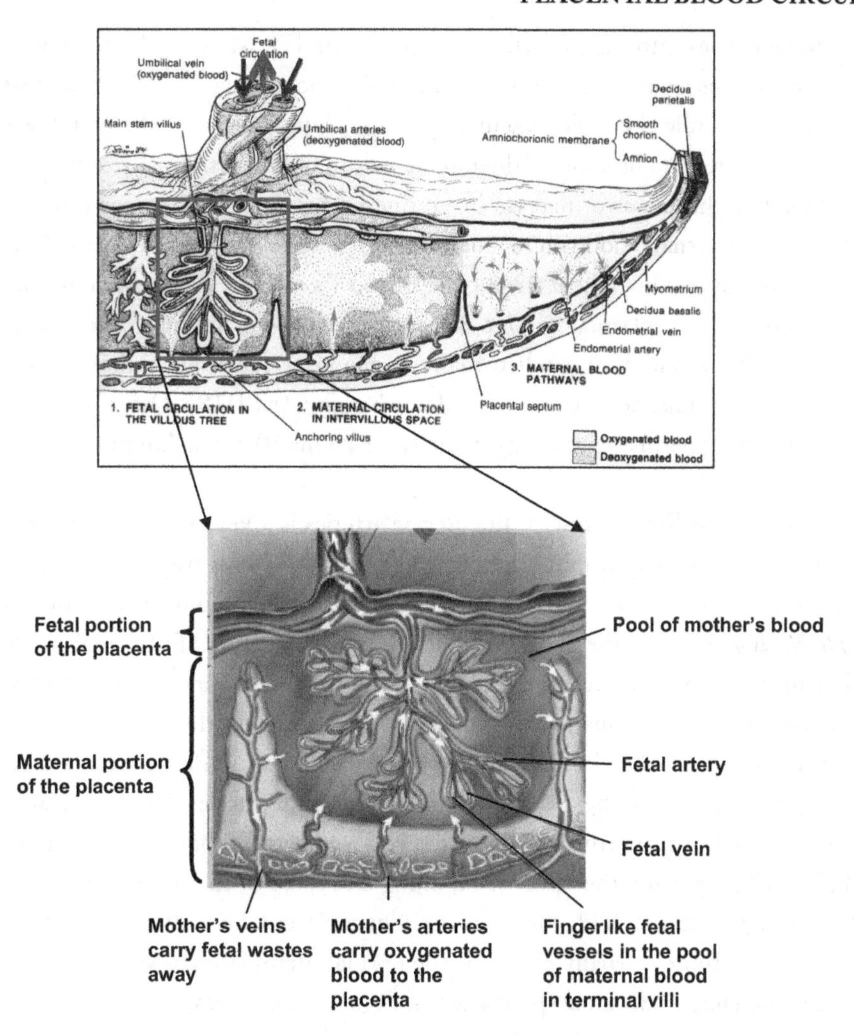

FIGURE 2.2: A schematic drawing of a section through a full-term placenta. Upper panel: (1) The relation of the villous chorion (C) to the decidua basalis (D) and the fetal-placental circulation; (2) The maternal blood flows into the intervillous spaces in funnel-shaped spurts, and exchanges occur with the fetal blood as the maternal blood flows around the villi; and (3) Maternal blood flow pathway. The inflowing arterial blood pushes venous blood into the endothelial veins, which are scattered over the entire surface of the decidua basalis. The cotyledons are separated from each other by the placental (decidual) septa. The umbilical cord vein carries oxygenated blood to the fetus and umbilical cord arteries carry deoxygenated blood to the placenta. Reproduced from 22nd Edition of Williams Obstetrics (page 65). Used with permission from McGraw-Hill. *Lower panel*: Fetal and maternal portions of the placenta. The arrows indicate the direction of the blood flow. Reproduced from *Williams Obstetrics* 22e with permission of McGraw-Hill Professional.

120 spiral arterial entries into the intervillous space at term [7]. Maternal blood traverses through the placenta intervillous space and drains back through venous orifices in the basal plate, then returns the maternal systemic circulation via uterine veins. Maternal-placental blood flow is propelled by maternal arterial pressure because of the unique nature of low-resistance uteroplacental vessels, which accommodate the massive increase in uterine perfusion over the course of gestation [7]. During pregnancy, maternal blood volume increases progressively from 6–8 weeks of gestation and reaches a maximum approximately at 32–34 weeks and then keeps relatively constant until term. In general, maternal blood (plasma) volume is increased up to 40–50% near term compared to the nonpregnant state. Gowland et al. studied maternal blood perfusion in human placenta from 20 weeks of gestational age until term using echo planar imaging (EPI) [8]. They found that in normal pregnancies the average perfusion rate was about 176 ± 24 ml/100 gram/minute.

Spiral artery remodeling: Remodeling of the uterine arteries is a key event in early pregnancy that begins after implantation. The trophoblast differentiates into villous trophoblasts and extravillous trophoblasts. These trophoblasts have distinct functions when in contact with maternal tissues. ***Villous trophoblasts*** give rise to the chorionic villi, the major structure of placental cotyledon. Chorionic villi primarily transport oxygen and nutrients between fetus and mother. ***Extravillous trophoblasts*** migrate into the decidua and myometrium and penetrate the maternal vasculature. The extravillous trophoblasts can be classified as interstitial trophoblasts and endovascular trophoblasts. Interstitial trophoblasts invade the decidua and surround spiral arteries. Endovascular trophoblasts invade spiral arteries. In the uterine spiral arteries, endovascular trophoblasts interdigitate between the endothelial cells, replacing the endothelial lining and most of the musculoelastic tissue in the vessel walls, thereby creating a high-flow, low-resistance placental circulation. "High flow and low resistance" is the description usually given for the normal uteroplacental vasculature as a result of physiological remodeling of decidual spiral arteries. Figure 2.3 illustrates the process of spiral artery remodeling during pregnancy.

Placental blood flow is increased throughout pregnancy. The volume of placental blood flow is about 600–700 ml/minute (80% of the uterine perfusion) at term. Steep falls in the pressure occur in the transition from uterine arteries to intervillous spaces. The pressure is about 80–100 mmHg in uterine arteries, 70 mmHg in spiral arteries, and only 10 mmHg within intervillous space. The low-resistance of uteroplacental vessels and the gradient of blood pressure between uterine arteries and placental intervillous space allow the maternal blood to perfuse the intervillous space efficiently and effectively. The blood in the intervillous space is therefore completely exchanged two to three times per minute. In general, the spiral arteries are perpendicular to the uterine wall, while the veins are parallel to the uterine wall. This arrangement facilitates closure of the veins during uterine contractions and prevents squeezing of maternal blood from the intervillous space.

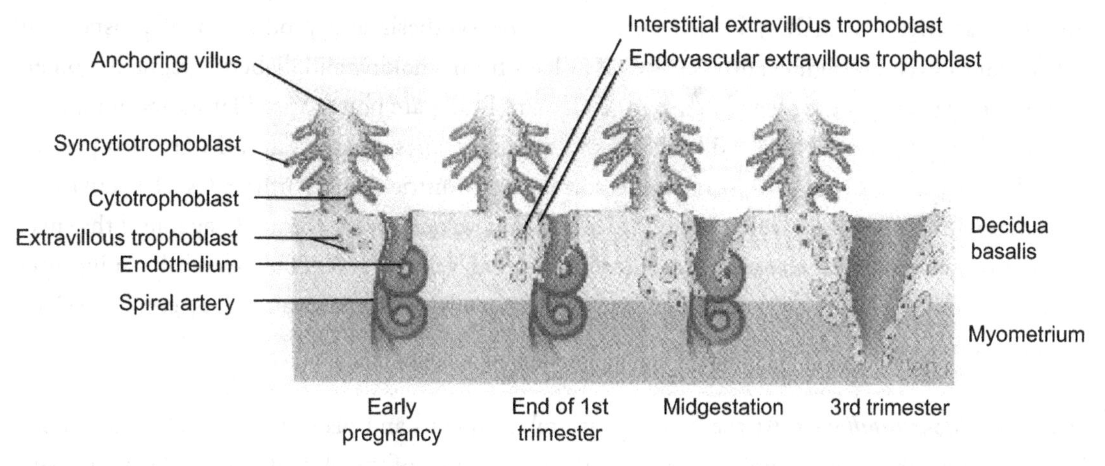

FIGURE 2.3: The process of spiral artery remodeling during pregnancy. In early pregnancy, two types of extravillous trophoblasts are found outside the villous, endovascular and interstitial trophoblasts. Endovascular trophoblasts invade and transform spiral arteries during pregnancy to create low-resistance blood flow that is characteristic of the placenta. Interstitial trophoblasts invade the decidua and surround spiral arteries. The figure is adapted from the 23rd Edition Williams Obstetrics, page 50 (Figure 3-12). Adapted from *Williams Obstetrics* 23e with permission of McGraw-Hill Professional.

2.2 FETAL-PLACENTAL CIRCULATION

Umbilical cord: The umbilical cord is the lifeline that attaches the placenta to the fetus. During prenatal development, the umbilical cord comes from the same zygote as the fetus. The umbilical cord in a full-term human neonate averages ~50–70 centimeters (20 inches) long and ~2 centimeters (0.75 inches) in diameter. It extends from the fetal umbilicus to the fetal surface of the placenta or chorionic plates. The cord is not directly connected to the mother's circulatory system. Instead it joins the placenta, which transfers materials to and from the mother's blood without allowing direct mixing. The umbilical cord contains one vein (the umbilical vein) and two arteries (the umbilical arteries) buried within Wharton's jelly. The umbilical vein carries oxygenated, nutrient-rich blood from the placenta to the fetus, and the umbilical arteries carry deoxygenated, nutrient-depleted blood from the fetus to the placenta (Figure 2.2). Any impairment in blood flow within the cord can be a catastrophic event for the fetus.

Umbilical vessels are sensitive to various vasoactivators, such as serotonin, angiotensin II, and oxytocin. The contractility of smooth muscles in vessel walls is also influenced by substances produced by the neighboring endothelial cells in a paracrine manner [9]. Umbilical cord vessels produce several potent vasodilators. For example, an *in vitro* study has shown that the endothelium from umbilical vein (HUVECs) produces far more prostaglandins than the endothelium from umbilical arteries (HUAECs) [10]. Interestingly, the synthesis and production of prostacyclin (PGI_2) and PGE_2 are significantly less by HUVECs from smoking and diabetic pregnant women than in normal pregnant women [11]. Both PGI_2 and PGE_2 are potent vasodilators and inhibitors for platelet aggregation. Nitric oxide (NO) and atrial natriuretic peptide (ANP) are also present in umbilical vessels. Giles et al. studied the correlation of nitric oxide synthase (NOS) activity in placentas with Doppler ultrasound umbilical artery flow velocity wave-forms. They found that placentas from women with abnormal umbilical artery flow velocity waveforms showed significantly lower mean NOS activity than did placentas from women with normal umbilical artery flow velocity wave-forms [12].

Placental villous capillaries: At the junction of umbilical cord and placenta, the umbilical arteries branch to form chorionic arteries and traverse the fetal surface of the placenta in the chorionic plate and branch further before they enter into the villi. The chorionic arteries are easily recognized because they always cross over the chorionic veins. These vessels are responsive to vasoactive substances as mentioned above. About two thirds of the chorionic arteries form networks supplying the cotyledons in a *pattern of disperse-type branching*. The rest of the chorionic arteries radiate to the edge of the

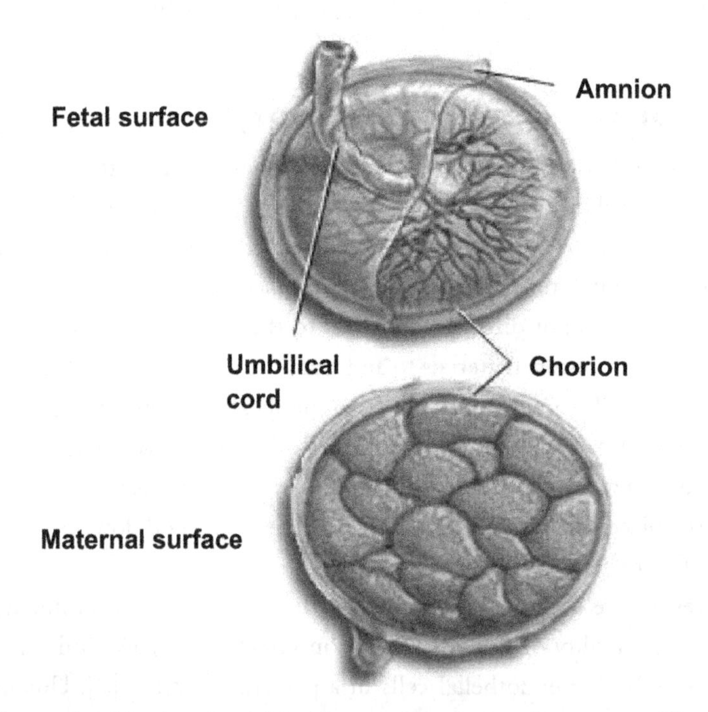

FIGURE 2.4: Maternal and fetal surface of the placenta. Please note the pattern of disperse-type branching of fetal vessels (fetal surface) in the chorionic plate. (From http://www.walgreens.com/marketing/library/graphics//images/en/17010.jpg). Used with permission of A.D.A.M.

placenta and down to a network. Figure 2.4 shows the maternal and fetal surfaces of a placenta; note the disperse-type branching pattern of fetal vessels (fetal surface) in the chorionic plate.

Each umbilical cord artery generally provides eight or more terminal chorionic plate arteries, which are referred to as stem arteries of the peripheral trunci chorii to the fetal villous cytyledons. The first order branches have an average length of 5–10 mm; the artery is an average of 1.5 mm in diameter with the accompanying vein being about 2 mm. These truncal vessels divide into four to eight horizontal cotyledonary vessels of the secondary order, with an average diameter of 1 mm. The horizontal distance varies with the size of the cotyledon, and as they curve toward the basal plate, they begin branching into the third-order villous branches. There are about 30–60 branches in each cotyledon, with calibers of 0.1–0.6 mm and lengths of 15–25 mm. In the villi, the third-order villous branches form an extensive arteriocapillary venous system, ***villous capillaries***, bringing the fetal blood extremely close to the maternal blood; but no intermingling between fetal and maternal blood occurs. There are about 15–28 cotyledons per placenta.

The villous capillaries are branches of the umbilical vessels, and the capillary networks are the functional unit of maternal-fetal exchange. The blood pressure in the umbilical arteries averages about 50 mmHg, and the blood flows through smaller vessels that penetrate the chorionic plate to the capillaries in the villi where arterial blood pressure falls to 30 mmHg. In the umbilical vein the pressure is 20 mmHg. The pressure in the fetal vessels and their villous branches is always greater than that within the intervillous space. This protects the fetal vessels against collapse.

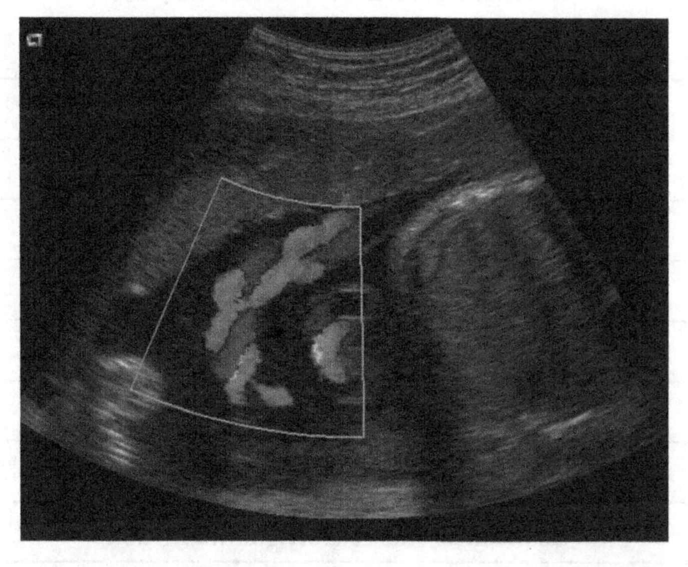

FIGURE 2.5: Ultrasound color imaging of umbilical cord arteries and vein. Red: umbilical cord vein; and blue: umbilical cord arteries. (from http://www.medical.siemens.com). Image courtesy of Siemens Healthcare.

Assessment of fetal blood flow: Ultrasound and Doppler flow measurements provide means to visualize the umbilical cord and to evaluate the fetal blood flow. Figure 2.5 shows an example of an ultrasound color image of umbilical cord arteries and vein. By measuring the amount of forward blood flow through the umbilical artery during both fetal systole and diastole, an overall measure of fetal health can be obtained. In general, the more forward blood flow from the fetus to the placenta through the umbilical artery, the healthier the fetus. Table 2.1 summarizes measurements and vessel blood flow characteristics accessed by ultrasound and Doppler devices. The mean absolute vein blood flow is about 443 ± 92 ml/min in normal umbilical cord between 24 and 29 weeks of gestation, and reduced absolute vein blood flow is associated with low fetal birthweight [13]. Therefore, an assessment of fetal blood flow through the umbilical cord by ultrasound color Doppler sonography has proven to be a valuable noninvasive procedure for assessing fetal well-being during pregnancy.

Abnormal insertion of umbilical cord: In more than 90% of placentas, the umbilical cord inserts on the fetal surface (chorionic plate) of the placenta more than 3 cm from the margin, less than 10% insert at or near the margin, and about 1% insert in the placental membranes. Chorionic plate arteries and veins branch from the umbilical cord insertion (Figure 2.6A). Arteries always cross veins. Recent studies have shown that peripheral umbilical cord insertion is associated with a decreased den-

TABLE 2.1: A list of parameters of umbilical cord measurements and vessel blood flow characteristics accessed by Ultrasound [13]
Cord length (cm)
Cord cross-sectional area (mm^2)
Artery cross-sectional area (mm^2)
Vein cross-sectional area (mm^2)
Coiling index
Wharton's jelly area (mm^2)
Artery pulsatility index
Absolute vein blood flow (ml/min)
Vein blood flow for fetal weight (ml/kg/min)
Vein blood flow mean velocity (cm/second)

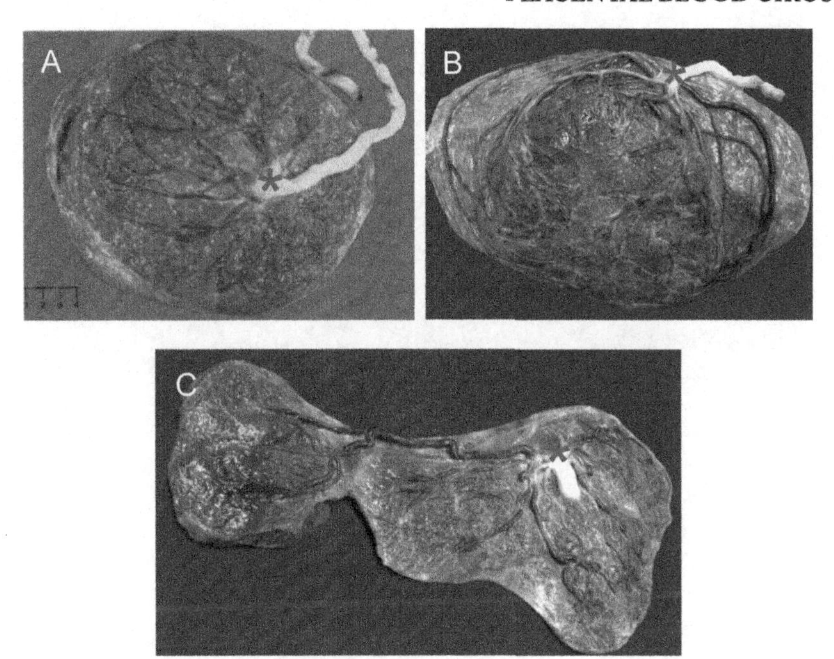

FIGURE 2.6: Chorionic plates arteries and veins branch from the umbilical cord insertion. A: Chorionic plate of a normal term placenta; B: umbilical cord insertion in the margin of the placenta; and C: extra lobe present with a velamentous vessel connecting. Asterisks show the cord insertion in the chorionic plate. (A: adapted from Kraus FT et al.; Placental Pathology, Figure 1–7; B and C adapted from Kaplan CG. J Clin Pathol 2008: 61: 1285-1295). A: Adapted from *Placental Pathology*, Kraus FT et al. with permission of the American Registry of Pathology. B and C: Adapted from Kaplan CG. *J Clin Pathol* 2008 with permission of the BMJ Group.

sity of chorionic plate vessels and an altered fetoplacental weight ratio, which may reflect deleterious effects on placental function and fetal growth. Examples of abnormal insertion of the umbilical cord include the cord inserts near the margin of the placenta (Figure 2.6B) or an extra lobe present with a velamentous vessel connecting (Figure 2.6C). Peripheral cord insertion, velamentous and marginal, is associated with increased frequency in abortions, preterm labor, and discordant fetal growth. The etiology of peripheral cord insertion is not clear, but it may result from malpositioning of the blastocyst at implantation, with consequent aberrant placental disk orientation, or with placental shift from its initial implantation site, leaving the cord insertion behind. Velamentous cord insertion is considered a marker of poor placentation with decreased chorionic and placental vascularization.

• • • •

CHAPTER 3

Structure of the Placenta

Villous "trees" are the main structure of the placenta. Based on the developmental stage, villous structure, vessel branches, histologic features, and vessel-cell type components, at least five types of villi have been described [14]. An illustration of the architecture of different villous trees is shown in Figure 3.1A.

(1) Stem villi. Stem villi connect to the chorionic plate and are characterized by a condensed fibrous stroma containing large vessels and microvessels. Stem villi develop vessels with a smooth muscle investment and central stromal fibrosis. The trophoblast layer of stem villi is partly replaced by fibrin-type fibrinoid as gestation proceeds. The function of stem villi is to support the structures of the villous trees. Because of the low degree of fetal capillary and degeneration changes of the trophoblasts, fetal-maternal exchange and endocrine activity of stem villi are usually negligible [14].

(2) Immature intermediate villi. Immature intermediate villi are bulbous, peripheral, and immature continuations of stem villi. This type of villi has a loose or reticular stroma and Hofbauer cells, more prominent vessels and a discontinuous cytotrophoblast layer. The outer syncytiotrophoblast layer remains continuous throughout development. Immature intermediate villi are considered the growth centers of the villous trees. Immature intermediate villi are probably the principal sites of exchange during the first and second trimesters, as long as terminal villi are not yet differentiated [14].

(3) Mature intermediate villi. Mature intermediate villi are long, slender, peripheral ramifications that lack fetal vessels in the stroma. Mature intermediate villi produce the terminal villi. The high degree of fetal vascularization and the large share in the exchange surface make them important for fetal-maternal exchange.

(4) Terminal villi. Terminal villi are linked to stem villi by intermediate structures. These villi are grape-like structures characterized by a high degree of capillarization and highly dilated sinusoids. In term placenta, terminal villi are smaller with less stroma and a discontinuous cytotrophoblast layer, and contain 4–6 fetal capillaries per cross section. The fetal capillaries of the villous core oppose against thin attenuated syncytiotrophoblasts forming vasculosyncytial membranes. In the terminal villi, the fetal capillary vessels and syncytiotrophoblasts are separated by only a thin basement membrane with a minimal mean maternal-fetal diffusion distance ~3.7 μm, which make

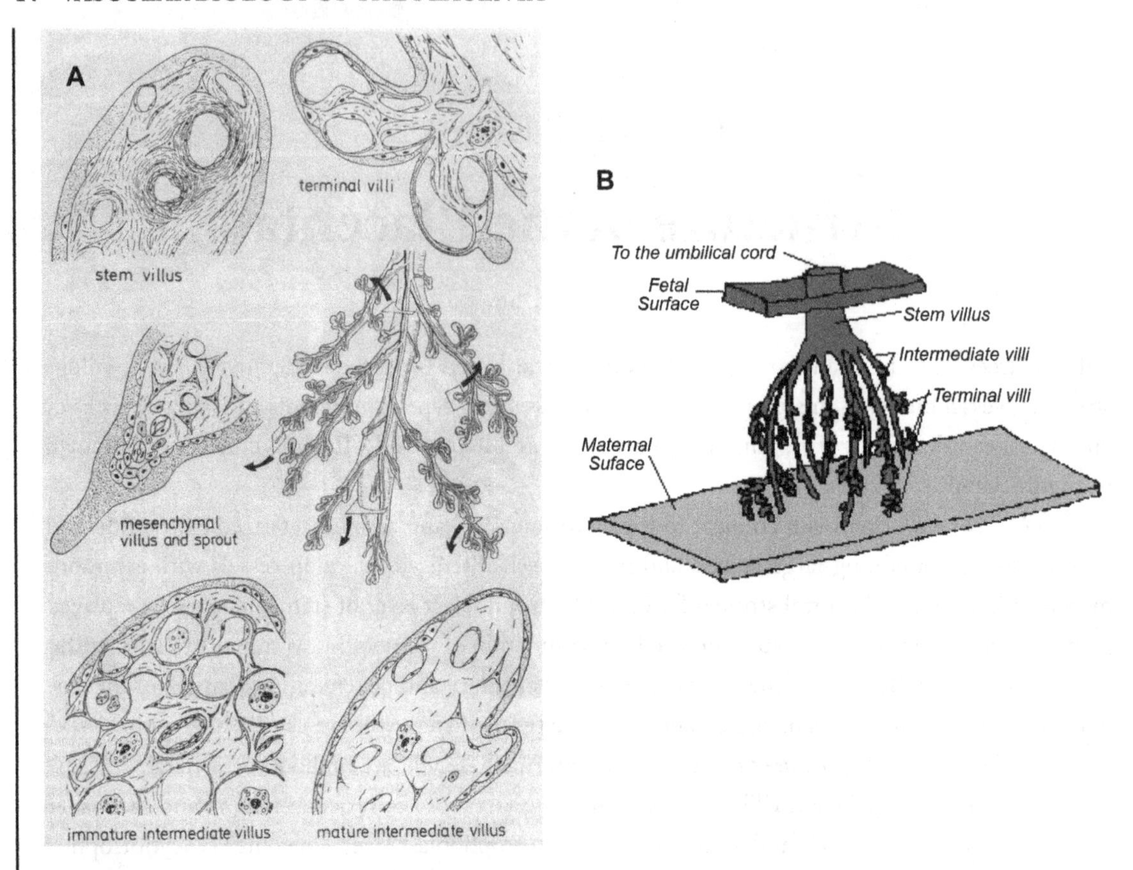

FIGURE 3.1: Architecture of different villous and vessel branches of a cotyledon. A: types of placental villi in human placenta. B: A villous tree connects to the fetal surface (chorionic plate) and the maternal surface (basal plate). The villous trees are so named because they resemble trees and their basic structure is established early in gestation. A "trunk" dives down into the placenta from the fetal surface vessels, as the stem villous, which divides into large branches, the intermediate villi, with nutrient gathering structures at their ends, followed by the terminal villi. However, during the second and third trimesters, the terminal villi continue to mature in such a way as to increase the quantity and quality of oxygen and nutrient exchange across the terminal villi between the maternal and fetal blood, in response to the demands of the growing fetus. (A is adapted from Benirschke et al. Fifth Edition of Pathology of the Human Placenta (page 123). Used with permission from Springer, and B is from http://showcase.netins.net/web/placenta/placentaltriage101.htm.). A: Reproduced from *Pathology of the Human Placenta*, Benirschke et al. with permission of Springer. B: Used with permission of DSM Pathworks Inc.

terminal villi the most appropriate place for diffusive exchange. In the normal mature placenta, the terminal villi comprise nearly 40% of the villous volume of the placenta. Because of their small diameters, the sum of their surfaces account for about 50% of the total villous surface and 60% of villous cross sections [14]. Terminal villi, the functional unit of the placenta, transfer electrolytes, O_2, CO_2 and nutrients between the mother and fetus.

(5) Mesenchymal villi. Mesenchymal villi are the most primitive type of villi during early stages of pregnancy. Mesenchymal villi have loose stroma, inconspicuous capillaries, and two complete surrounding trophoblast layers, a cytotrophoblast layer surrounding the villous core, and an outer syncytiotrophoblast on the villous surface. Fetal capillaries are poorly developed and never show sinusoidal dilatation. The unvascularized tips of mesenchymal villi are referred to as villous sprouts (Figure 3.2). The function of mesenchymal villi is very important during the first few weeks of pregnancy. Mesenchymal villi are the place of villous proliferation and they perform almost all endocrine activities. With advancement of pregnancy, their primary function is to sustain villous growth. At term, their share in total villous volume is less than 1% [14].

Placenta villous development starts with mesenchymal villi. Up to 5 weeks postconception (*p.c.*), all placental villi are of the 'mesenchymal' type (containing trophoblast and villous sprouts) [15]. Mesenchymal cells later invade these villi forming secondary villi (immature/intermediate villi) and also giving rise to placental blood vessels. Trophoblast syncytiolization leads to the formation of villous sprouts. Mesenchymal villi are continuously formed throughout pregnancy, but dominate during the first and second trimesters [15]. Villous sprouts further transform into immature/mature intermediate villi then to terminal villi [15,16]. Trophoblast sprouting, proliferation, and formation of finger-like trophoblast protrusions lead to mesenchymal invasion and local fetal angiogenesis

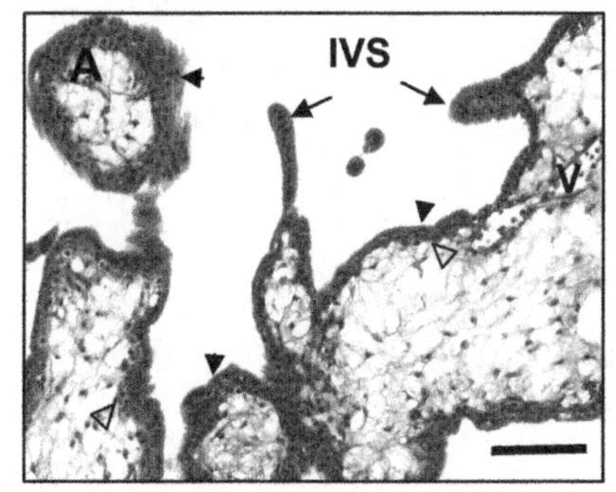

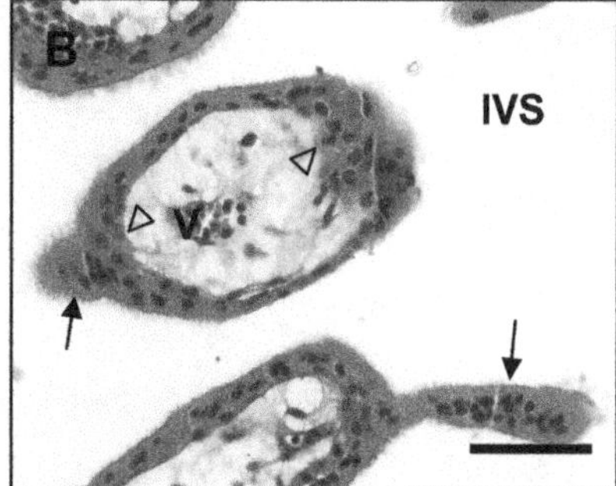

FIGURE 3.2: Mesenchymal villi and villous sprouts in first-trimester placentas. Open arrowhead: cytotrophoblasts; solid arrowhead: syncytiotrophoblasts; and arrow: villous sprouts; V: fetal vessel; and IVS: intervillous space, respectively. A: bar = 100 micron, and B: bar = 50 micron.

[16]. Villous core fetal vessel formation and fetal-placental blood flow begins approximately around 6–8 weeks *p.c.*

The weight of the placenta is about 20 grams at 10 weeks of gestation and 150–170 grams at 20 weeks of gestation. A mature placenta weighs about 500–600 grams and consists of 15–28 "cotyledons." The stem villus is the major structural unit of the fetal cotyledon. Each cotyledon begins with a stem villus that divides into 3-5 immature/mature intermediate villi, which further branches into 10–12 terminal villi (Figure 3.1B). Some terminal villi float freely in the intervillous space, whereas others are attached to the decidua, providing structural stability for the placenta.

• • • •

CHAPTER 4

Cell Types of the Placenta

Placenta villi are composed of three layers of components with different cell types in each: (1) syncytiotrophoblasts/cytotrophoblasts that cover the entire surface of the villous tree and bathe in maternal blood within the intervillous space; (2) mesenchymal cells, mesenchymal derived macrophages (Hofbauer cells), and fibroblasts that are located within villous core stroma between trophoblasts and fetal vessels. Hofbauer cells synthesize VEGF and other proangiogenic factors that initiate vasculogenesis in the placenta; and (3) fetal vascular cells that include vascular smooth muscle cells, perivascular cells (pericytes), and endothelial cells.

4.1 CYTOTROPHOBLASTS AND SYNCYTIOTROPHOBLASTS

Trophoblasts (from Greek to feed: *threphein*) are cells forming the outer layer of a blastocyst, which provides nutrients to the embryo, and develops into a large part of the placenta. They are formed during the first stage of pregnancy and are the first cells to differentiate from the fertilized egg. Villous trophoblasts have two cell populations: undifferentiated cytotrophoblasts and fully differentiated syncytiotrophoblasts. The syncytiotrophoblasts are a continuous, specialized layer of epithelial cells. They cover the entire surface of villous trees and are in direct contact with maternal blood. The surface area of syncytiotrophoblasts is about 5 square meters at 28 weeks' gestation and reaches up to 11–12 square meters at term [1]. Figure 4.1 illustrates total placental surface areas at different gestational ages. Fully developed terminal villi are the functional unit of maternal-fetal oxygen exchange and nutrient transport. Figure 4.2 shows hematoxylin and eosin (H & E) staining of a cross section of terminal villi of a term placenta and an electron microscopic section of a terminal villous. Note the relation of syncytiotrophoblasts and fetal capillaries. The fetal capillary basement membrane is very close to the maternal blood in the intervillous space.

In terms of fetal-maternal communication, it is mainly the syncytialized trophoblasts that orchestrate the complex biomolecular interactions between the fetus and mother. Not only do placental trophoblasts provide structural and biochemical barriers between the maternal and fetal compartments during pregnancy, they also serve as an important endocrine organ that produces numerous growth factors and hormones that support and regulate placental and fetal development and growth [17–20]. Table 4.1 is a list of major hormones produced by placental

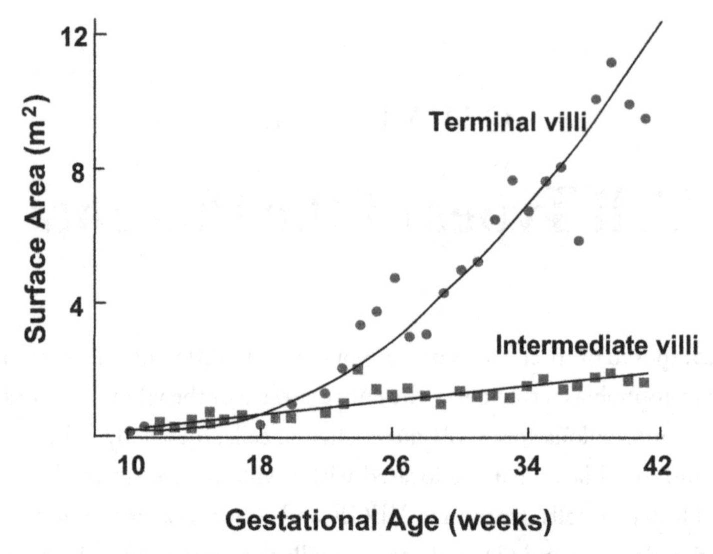

FIGURE 4.1: Placental surface areas at different gestational ages. (■) areas of intermediate villi; (●) areas of terminal villi. (Adapted from The Physiology of the Human Placenta, by Page K, Figure 2.7, published by UCL press). Reproduced from *The Physiology of the Human Placenta*, Page K with permission of Taylor & Francis.

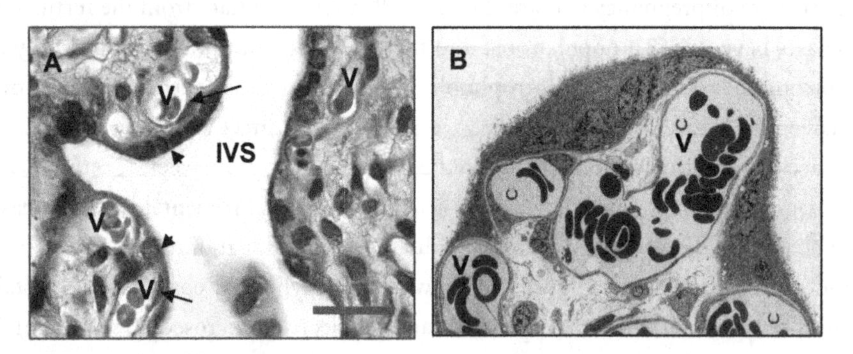

FIGURE 4.2: Hematoxylin and eosin (H & E) staining of a terminal villi tissue section and an electron microscopic section of a terminal villus. Fetal red blood cells are seen in villous core fetal vessels. Please note the intimate relationship of syncytiotrophoblasts and fetal vessels in the terminal villi. A: from a 38-week placenta, bar = 25 microns. Arrow: villous core fetal vessel endothelial cells; arrowhead: syncytiotrophoblasts; and V: villous core fetal vessels. B: transmission electronic microscopic section of a terminal villous. Adapted from Benirschke et al. Fifth Edition of Pathology of the Human Placenta (page 57, Figure 6.6). Used with permission from Springer. Adapted from *Pathology of the Human Placenta*, Benirschke et al. with permission of Springer.

TABLE 4.1: Function of major hormones produced by placental syncytiotrophoblasts
Human chorionic gonadotropin (hCG)
Synthesis of hCG begins before implantation, and is responsible for maintaining the maternal corpus luteum that secretes progesterone and estrogens. The level of hCG is the basis for early pregnancy tests. Production peaks at eight weeks and then gradually declines.
Estrogens and progesterone
The placenta produces progesterone independently from cholesterol precursors, and estrogen in concert with the fetal adrenal gland. By the end of the first trimester, the placenta produces enough of these steroids to maintain the pregnancy and the corpus luteum is no longer needed.
Human placental lactogen (hPL) or **human chorionic somatomammotropin**
(hCS) hPL is similar to growth hormone and influences growth, maternal mammary duct proliferation, and lipid and carbohydrate metabolism.
Human placental growth hormone
This hormone differs from pituitary growth hormone (GH) by 13 amino acids. From 15 weeks until the end of pregnancy, this hormone gradually replaces maternal pituitary GH. Its major function is the regulation of maternal blood glucose levels so that the fetus is ensured of an adequate nutrient supply.
Insulin-like growth factors
Stimulates proliferation and differentiation of the cytotrophoblasts.
Endothelial growth factor
Start to produce this hormone by 4 -5-week *p.c.*; It stimulates proliferation of trophoblasts.

syncytiotrophoblasts. The function of trophoblasts is tightly regulated by locally produced growth factors, components of the extracellular matrix (ECM), and binding between growth factors and proteoglycans. Furthermore, a successful pregnancy also largely depends on the angiogenic function of the trophoblast itself, i.e., its ability to invade maternal myometrial spiral arteries (during the first trimester) and the capacity to generate growth factors that vascularize the placenta during its mid- to later development (second and early third trimester). In terms of vasculogenesis and angiogenesis, it is believed that vascular endothelial growth factor (VEGF), placental growth factor (PlGF), fibroblast growth factors (FGFs), and their receptor families are key factors regulating trophoblast survival and angiogenesis in the placenta. It is well known that placental trophoblasts are the source of VEGF and PlGF. (See Chapter 7 for more details.)

Beneath the syncytiotrophoblasts are the cytotrophoblasts (Figure 3.2). These cells are considered to be stem cells for syncytiotrophoblasts. Cytotrophoblasts continually differentiate into syncytiotrophoblasts during villous formation and development. Cytotrophoblast invasion into the uterine spiral arteries is accompanied by loss of the endothelial lining and musculoelastic tissue in these vessels. This process of invasion is necessary for placental vascular remodeling in the early stages of the implantation process. As a result of this invasion process, there is a loss of elasticity and an increase in the luminal diameter of the spiral arteries. Consequently, the spiral arteries become low resistance vascular channels (Figure 2.3). Normal decidual vascular architecture formation is required to meet the gestational increases in the demand for blood flow to the placenta.

Preeclampsia is associated with shallow cytotrophoblast invasion: Preeclampsia is a hypertensive and multiple system disorder unique to human pregnancy. Preeclampsia is diagnosed with newly developed maternal hypertension and positive proteinuria after 20 weeks of gestation. Although the etiology of preeclampsia is unknown, evidence supports the notion that preeclampsia is associated with shallow cytotrophoblast invasion. Brosen et al. first described the abnormal "shallow" cytotrophoblast invasion in placentas from women whose pregnancies are complicated by preeclampsia [21]. They found that cytotrophoblast invasion of the uterus is only superficial, and the endovascular invasion does not proceed beyond the terminal portions of the spiral arterioles. The process of trophoblast invasion is normally completed by 20 to 22 weeks of gestation in normal pregnancy. However, in cases of preeclampsia it has been found that cytotrophoblast invasion of the uterine spiral arterioles is often incomplete by this time and spiral arteries fail to lose their muscular elastic components [21–23] Therefore, a critical underlying lesion in preeclampsia is the failure of extravillous trophoblasts to invade the muscular spiral arteries into their myometrial portion and convert them to "low-resistance" capacitance vessels. Consequently, placental hypoxia and reduced placental perfusion characterized with "low flow and high resistance" are the central hallmarks of preeclampsia. Figure 4.3 illustrates abnormal spiral artery remodeling that results in low flow and high resistance in cases of preeclamptic placenta [24].

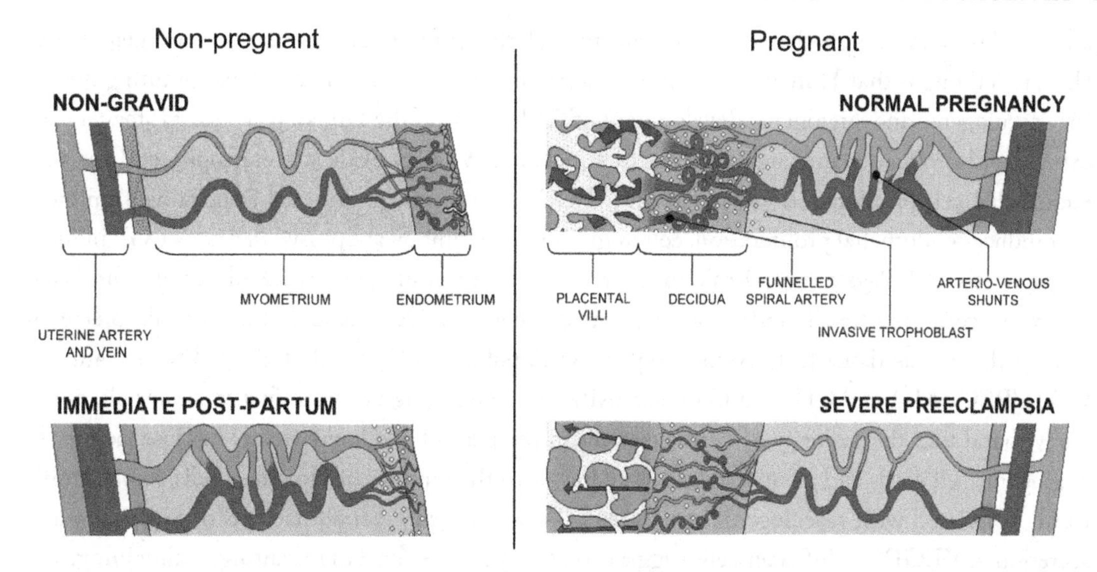

FIGURE 4.3: Illustration of uterine and placental vasculature in the non-pregnant, pregnant and immediate post-partum state. Normal pregnancy is characterized by the formation of large arterio-venous shunts that persist in the immediate post-partum period. By contrast pregnancies complicated by severe preeclampsia are characterized by minimal arterio-venous shunts, and thus narrower uterine arteries characterized with "low flow and high resistance." Red shading = arterial; blue shading = venous. Adapted from Burton et al. Placenta 2009; 30 (6), 473-482. Adapted from *Placenta*, Burton et al. 2009 with permission of Elsevier.

4.2 VILLOUS CORE STROMA CELLS

Demir et al. analyzed the ultrastructure of human placental villous tissues from 21 days *p.c.* through 40 weeks of gestation [25]. They found that early villous core fetal cells at 21 days *p.c.* appear to originate from macrophage-like mesenchymal precursor cells—mesenchymal stem cells. These cells are differentiated by vasculogenesis and angiogenesis. In the early stage of placental development, mesenchymal stem cells transform into hemangioblastic cell cords, which are believed to be the precursors of capillary endothelial cells and hematopoietic stem cells [25]. Mesenchymal stem cells can also differentiate into perivascular cells, which are considered predecessors of capillary endothelial cells. In the late stage of gestation, fetal villous core angiogenesis results from the proliferation of existing endothelial cells and pericytes rather than via the expansion of hemangioblastic cells [25].

Hofbauer cells were first described by Hofbauer in 1903 [26]. Hofbauer cells are the macrophages in the placenta villous stroma. These cells are of mesenchymal origin and expand during the first and second trimesters in placental villous tissues. Hofbauer cells are antigen-presenting cells in the placenta, which play a critical role in maintaining host defense. These placental macrophages also contribute to trophoblast differentiation and angiogenesis by producing various growth factors and cytokines. Although the exact mechanism underlying placental vascular development is not clear, it is thought that Hofbauer cells must govern mesenchymal villi and villous sprouting during the vasculogenic and angiogenic development of fetal vessels in the human placenta. Hofbauer cells express and produce many angiogenic factors including VEGF, FGF, vasculotropin, and vascular endothelial cell proliferation factor [27–30]. VEGF is a highly specific endothelial cell mitogen that induces endothelial proliferation, cell migration, and inhibits apoptosis. *In vivo* VEGF induces angiogenesis and plays a central role in the regulation of vasculogenesis. Demir et al. found that Hofbauer cells are often found close to cytotrophoblasts, vasculogenic and angiogenic precursor cells in the villous tissue [30]. Similar reports were published by Seval et al. [31] who examined CD31/PECAM-1 and CD68 immunoreactivities by staining tissues from first trimester placentas. They found that the majority of Hofbauer cells in the placental villous core are in close vicinity to angiogenic cell cords and primitive vascular tubes, with the number of Hofbauer cells being significantly correlated with vasculogenic structures [31]. The developmental patterns of expression and secretion of VEGF by Hofbauer cells suggests that they are involved in recruiting, maintaining, and forming angiogenic cells during the course of villous core development. These findings further support the idea that molecular interactions between Hofbauer-cell derived VEGF with its receptors VEGFR-1 (Flt-1) or VEGFR-2 (Flk-1/KDR) on vascular precursor target cells regulate vasculogenesis and angiogenesis in the placenta [30].

Hofbauer cells also express sprouty (*Spry*) proteins and Spry interaction protein c-Cbl [32]. Among the four isoforms of *Spry* proteins, Spry-2 has been found to be involved in the FGF signaling, modulating different types of tissue branching during placental development. Spry-2 expression was abundant in the Hofbauer cells first trimester placental tissues, localizing near the cytotrophoblasts and syncytiotrophoblasts [32]. Spry proteins are important regulators of branching morphogenesis and growth-factor signaling. Although Spry proteins have been shown to direct tubular morphogenesis associated with tracheal/lung development, uteretic budding, and endothelial angiogenesis [33], however, a reduction of Spry2 expression by siRNA in villous explant culture paradoxically shows a marked increase in villous outgrowth [34]. Although their roles in trophoblast sprouting have not yet been defined, Hofbauer cells clearly modulate trophoblast migration and branching during chorionic villous tree development [32].

4.3 PERICYTES/ENDOTHELIAL CELLS

Pericytes are perivascular cells with dendritic processes that surround capillary endothelium and venules. The ability of pericytes to support endothelial cells is fundamental for maintaining vessel stability and microvascular integrity [35] by restricting growth-factor-induced endothelial cell migration and proliferation [36,37]. Pericytes have also been shown to influence blood vessel functions including microvascular contractility, solute permeability, and smooth muscle stem cell function.

In the placenta, pericyte function is poorly understood. However, it is believed that pericytes are derived from mesenchymal stem cells and that they participate in different phases of placental vasculogenesis. Challier et al. made several observations [37]. When pericytes are co-cultured with endothelial cells, pericytes rapidly become the dominant cell type, indicating higher proliferative potential of pericytes than endothelial cells. Both pericytes and endothelial cells are involved in the formation of vascular "nodules" seen in prolonged in vitro culture. Depending on their functional state, degree of differentiation, and spatial relationship with endothelial cells and trophoblasts, pericytes play diverse roles during placental vascular development and angiogenesis. Although signaling mechanisms between pericytes and endothelial cells and between pericytes and trophoblasts are largely unknown, it is believed that pericytes play critical roles in maintaining microvessel stability and integrity, especially in the terminal villi, which are characterized by their high degree of capillarization and the presence of highly dilated sinusoids. Their intimate relationship with endothelial cells makes them likely candidates for sensing and regulating blood flow in the terminal villi of the placenta.

Placental microvascular endothelial cells exhibit remarkable changes of their angiogenic status throughout pregnancy, i.e., dominant angiogenic activity is associated with rapid placental expansion during the first and second trimesters, followed by a dominant angiostatic condition that is associated with placental growth arrest when pregnancy approaching to term. Thus, adequate placental microvascular endothelial function is key for a successful pregnancy. The inherent angiogenic regulation of these cells, their complex interactions with perivascular cells, and the appropriate vascularization responses require endothelial cells to rapidly remodel and adapt, paralleling the special needs of pregnancy, all of which are genetically programmed. Most studies that attempt to elucidate the functional role of placental microvascular endothelial cells have been restricted to microscopic analysis of frozen tissue sections [38]. Although it has been reported that placental microvascular endothelial cells express VEGF and its receptors, integrins, thrombomodulin, adhesion molecules, and HLA-B/C [38], an understanding of the molecular and functional regulation of these cells during placental development remains limited.

Lang et al. isolated endothelial cells from fetal artery and vein of human placenta, termed HPAECs and HPVECs, respectively [39]. Contrasting differences were found between HPAECs and HPVECs in terms of phenotype, genotype, and function. HPAECs are polygonal in shape with a smooth surface growing in loose arrangements and forming monolayers with classical endothelial "cobblestone" morphology. HPAEC express more artery-related genes such as hey-2 (hairy/enhancer-of-split related with YRPW motif, 2), connexin 40, and depp (decidual protein induced by progesterone) and more endothelial-associated genes than HPVECs. In comparison, HPVECs are spindle-shaped cells with numerous microvilli at their surface. These cells grow closely against each other, forming fibroblastoid "swirling" morphology at confluence with shorter generation and population doubling times than HPAECs [39]. Compared to HPAECs, HPVECs express more development-associated genes including gremlin, mesenchymal homeobox 2, and stem cell protein DSC54. These cells also show an enhanced potential to differentiate into adipocytes and osteoblasts compared to HPAECs [39]. In addition, HPAECs have a higher proliferative response to VEGF stimulation, whereas HPVECs are more responsive to PlGF [39]. Table 4.2 shows a list of endothelial-associated genes and their respective function in HPAECs and HPAECs [39].

TABLE 4.2: Endothelial-associated genes in human placental artery and vein endothelial cells

GENE NAME	FULL NAME	FUNCTION
	HPAECs:	
Hey-2	Hairy/enhancer-of-split related with YRPW motif 2	Locate in the nucleus to interact with a histone deacetylase to repress transcription
Connexin40	Gap junction protein, delta 4 (41kDa)	Involved in the formation of gap junction and intracellular conduits that connect the cytoplasm of connecting cells
Connexin37	Gap junction protein, delta 4 (37 Kda)	Component of gap junction, provide channels for the diffusion of low molecular weight material from cell to cell
MGP	Matrix Gla protein	An inhibitor of arterial wall and cartilage calcification
NOS3	Nitric oxide synthase 3 (endothelial cells)	Synthesis of nitric oxide
Depp	Chromosome 10 open reading frame 10	Production of this gene leads to phosphorylation and activation of transcription factor ELK1
Angpt2	Angiopoietin 2	Antagonist of angiopoietin 1 (ANGPT1) and endothelial TEK tyrosine kinase (TIE-2, TEK)
ESM1	Endothelial cell-specific molecule 1	Might play a role in endothelium-dependent pathological disorder
Ephin B1	Ephin ligand B1	Transduces signals to activate integrin-mediated migration, attachment and angiogenesis

TABLE 4.2: (*continued*)

GENE NAME	FULL NAME	FUNCTION
EphinB2	Ephin ligand B2	Involved in establishing arterial versus venous identity and perhaps in anastamosing arterial and venous vessels at their junctions
Flt-1	Fms-related tyrosine kinase 1 (VEGF-R1)	Binding to VEGF-A, VEGF-B and PlGF and playing important role in vasculogenesis and angiogenesis
KDR/flk-1	Kinase insert domain receptor (VEGF-R2)	Main mediator of VEGF-induced endothelial proliferation, survival, migration, tubular morphogenesis and sprouting
Edg1	Sphingosine-1-phosphate receptor	Involved in endothelial differentiation and activation, and cell-cell adhesion
Notch 4	Notch homolog	Cell fate decision
MDG1	DnaJ (Hsp40) homolog, subfamily B, member	Promote the turnover of unfolded surfactant protein C (unfolded protein response)
HPVECs:		
Gremlin	Gremlin	This gene encodes a member of the BMP antagonist family. It regulates organogenesis, body patterning and tissue differentiation
MEOX2	Mesenchymal homeobox 2	Play essential role in embryogenesis, development, and cell differentiation, transcription and DNA regulation
DSC54	Stem cell protein DSC54	A mesenchymal stem cell marker
Nrp-1	Neuropilin-1	Plays versatile role in angiogenesis, cell survival, migration and invasion
Nrp-2	Neuropilin 2	Interact with VEGF and play a role in vascular development

TABLE 4.2: (*continued*)

GENE NAME	FULL NAME	FUNCTION
HPVECs:		
EDG2	Lysophosphatidic acid receptor 1	Member of G protein-grouped receptor superfamily, mediates diverse biologic functions, including proliferation, platelet aggregation, smooth muscle cell contraction, and chemotaxis
Endoglin	Endoglin	A major glycoprotein on endothelial cells. It binds to TGFβ1 and TGFβ3 with high affinity
EPAS1	endothelial PAS domain protein 1	Encodes a transcription factor involved in induction genes regulated by oxygen

• • • •

CHAPTER 5

Oxygen Tension and Placental Vascular Development

5.1 OXYGEN TENSION AND ESTABLISHMENT OF MATERNAL-PLACENTAL CIRCULATION

The partial oxygen pressure is significantly different between the placenta and endometrium from 8 to 12 weeks of gestation [40]. During 8–10 weeks of gestation the partial oxygen pressure is significantly lower in the placental bed (PO_2 = 17.9 mmHg) than in the endometrium (PO_2 = 39.6 mmHg). The partial oxygen pressure is increased in the placental bed at the end of the first trimester and beginning of the second trimester. By 12–13 weeks of gestation, partial oxygen pressure is relatively similar between the placental bed and the endometrium [40]. Jauniaux et al. measured intrauterine gases and acid-base gradients inside the human fetoplacental unit at 7 to 16 weeks' gestation [41]. Respiratory gases and acid-base values were recorded by means of a multiparameter sensor, and samples were taken from inside the exocoelomic and amniotic cavities, the placental tissue, decidua, and fetal blood. It was found that before 11 weeks' gestation, placental PO_2 was 2.5 times lower than decidual PO_2 [41]. The low partial oxygen pressure in the placental bed before 11–12 weeks of gestation indicates that (1) early placental villous development occurs in an environment of relative hypoxia; and (2) the maternal-placental circulation is only partially established at the end of the first trimester. This concept is supported by a histologic study of decidual spiral arteries with trophoblast plugs in the vessels and the presence of maternal erythrocytes in the intervillous space (5). During the first 10–12 weeks of gestation, extravillous trophoblast "plugs" block the spiral arteries and prevent maternal blood flow entering the intervillous space, thereby creating an environment of physiological hypoxia in which placental and fetal developments take place [42]. These studies also imply that the end of the first trimester between 10 and 12 weeks of gestation is the critical period for establishing the maternal-placenta circulation, i.e., maternal blood flow into the intervillous space. These observations support the current accepted paradigm that "the first wave of endovascular trophoblast invasion into the decidual segments of spiral arteries occurs at 8 to 10 weeks of gestation."

As the placenta progresses into the second trimester, intervillous blood flow is initiated. Maternal blood flow begins to perfuse the intervillous space and PO_2 increases. These changes trigger trophoblasts switching from a proliferative to an invasive state, leading to "the second wave of invasion" occurring at 16–18 weeks of gestation. This second wave of trophoblast invasion results in the critical phase of trophoblast invasion that leads to the maternal myometrial spiral artery transformation into "low resistance" vessels, which establishes a high flow and low resistance utero-placental circuit. This physiological transformation is characterized by a gradual loss of the normal musculoelastic structure of the arterial wall and replacement by amorphous "fibrinoid" material in which trophoblast cells are embedded [43]. All of these physiological transitions are required for a successful pregnancy. The failure of trophoblast invasion and spiral artery transformation can be seen in cases of complicated pregnancies such as those involving preeclampsia and intrauterine fetal growth restriction (IUGR).

5.2 HYPOXIA AND TROPHOBLAST DIFFERENTIATION

As discussed before, early placental villous development occurs in an environment of relative hypoxia, known to be a stimulus of cytotrophoblast proliferation [44,45]. Fisher's group has intensively studied the proliferation and differentiation of placental trophoblasts. Using an *in vitro* organ culture model, they examined the effects of different O_2 tensions on cytotrophoblast proliferation and differentiation during the invasion process, and found that hypoxia blocks cytotrophoblasts from differentiating into cells with an "invasive" phenotype [46]. By culturing villous explants placentas that are from 6–8 weeks PC on Matrigel in either normal O_2 (20%) or reduced O_2 (2%) conditions, they notice that both integrin $\alpha 1$ and HLA-G expressions (two markers of invasive phenotype) in cytotrophoblasts were upregulated when villous explants were cultured in 20% O_2. In contrast, cytotrophoblasts cultured in 2% O_2 failed to express integrin $\alpha 1$ [45]. These data suggest that at least some aspects of cytotrophoblast differentiation/invasion are arrested by hypoxia. Zhou et al. also reported that vascular endothelial (VE)-cadherin, an endothelial junction molecule, may help regulate the invasion process by allowing the cytotrophoblasts to "mimic" the adhesion molecule expression associated with endothelial cells [47]. However, during experimental hypoxia, VE-cadherin is not expressed by cytotrophoblasts *in vitro*. This finding is supported by an apparent lack of VE-cadherin expression in cytotrophoblasts in the placenta from cases of preeclampsia, which is presumably hypoxic [44]. As a result of these events, the failure to remodel the decidual spiral arterioles would limit placental vascular development, reduce the placental blood supply, and increase placental vascular resistance. Consequently, these events promote a further sustained hypoxic environment. Therefore, it is believed that in preeclampsia, abnormally shallow cytotrophoblast invasion and faulty differentiation lead to inadequate vascular transformation [45,47]. These observations come to the hypothesis that abnormal placental adaptation or insufficient trophoblast function contributes to the pathogenesis of preeclampsia.

5.3 MODELS OF PLACENTAL HYPOXIA

During pregnancy, oxygen tension/content has a direct impact on trophoblast differentiation and proliferation and villous development. Based on the analyses of typical placental villous structural changes associated with clinical applications, Kingdom and Kaufmann proposed three models of placental hypoxia: preplacental hypoxia, uteroplacental hypoxia, and postplacental hypoxia [48].

Preplacental hypoxia: Preplacental hypoxia occurs in the condition of reduced oxygen tension or content within the maternal blood, which causes placental hypoxia such as pregnancy in women living at high altitude or maternal anemia. In response to relative long-term maternal hypoxia, increased trophoblast proliferation and increased villous capillary branching are the typical findings in the placenta.

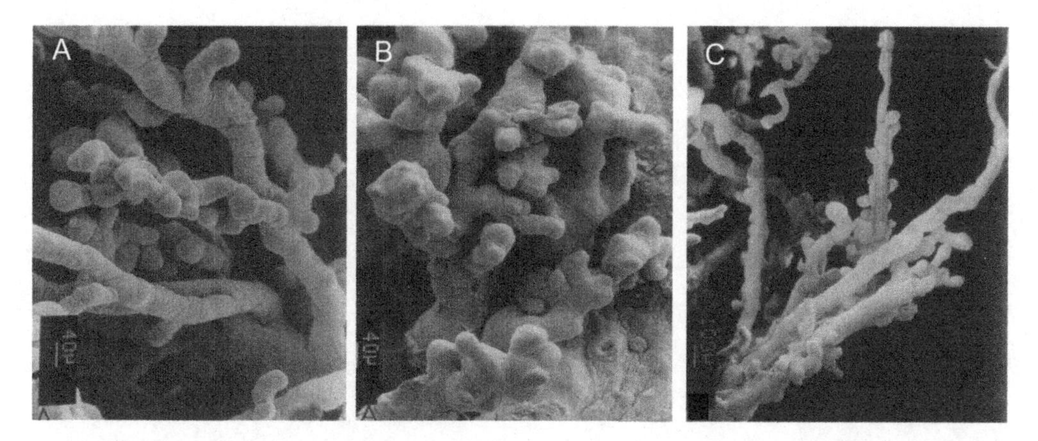

FIGURE 5.1: Scanning electron microscopy of placental villi from normal, preeclampsia and IUGR pregnancies. A: from a normal placenta: long slender, slightly curved mature intermediate villi with moderate number of grape-like terminal villi are the dominant features in the normal placenta.; B: from a preeclamptic placenta (uteroplacental hypoxia) showing the multiply branched, shot capillary loops with shot, knob-like indented villous surfaces; and C: from an IUGR placenta (postplacental hypoxia) showing long and poorly branched bundles of terminal villi. (The EM photos are adapted From Pathology of the human placenta, 5th Edition. Chapter 15: classification of villous maldevelopment. Edited by: Benirschke et al. A: from Figure 15.11A; B: from Figure 15.16 A; and C: from Figure 15.17 A). Reproduced from *Pathology of the Human Placenta* 5e, Benirschke et al. with permission of Springer.

Uteroplacental hypoxia: Uteroplacental hypoxia is caused by limited blood flow into the placenta, generally due to failed trophoblast invasion and maternal spiral artery remodeling. Predominant branching and short capillary loops are the feature of uteroplacental hypoxia. This is often seen in cases of IUGR with preserved end-diastolic umbilical flow in the third trimester, with or without late-onset preeclampsia [49].

Postplacental hypoxia: Postplacental hypoxia occurs when normally oxygenated blood reaches the intervillous space, but fetoplacental perfusion is compromised. Deficiency of terminal villi with long and poorly branched bundles of terminal villi is characterized in postplacental hypoxia [49]. This is seen in cases of IUGR combined with absent or reverse with end-diastolic umbilical flow in the second trimester with or without early-onset preeclampsia [49].

Figure 5.1 shows examples of scanning electron microscopic images of normal (A), pre-eclamptic (uteroplacental hypoxia B), and IUGR (postplacental hypoxia C) placentas.

• • • •

CHAPTER 6

Vasculogenesis and Angiogenesis of Human Placenta

Two distinct processes give rise to blood vessels: vasculogenesis and angiogenesis (3). There are three stages in placental vasculogenesis and angiogenesis during villous maturation (Figure 6.1) [50,51].

1. *Step-1:* Vasculogenesis starts with haemangiogenic stem-cell-induced cytotrophoblast differentiation that is regulated by VEGF in a paracrine manner.
2. *Step-2:* Angiogenesis I (activation) is a prevascular network formation process. Growth factors produced by cytotrophoblast cells and Hofbauer cells play a major role.
3. *Step-3:* Angiogenesis II (remodeling) is a process of differentiation of perivascular cells (pericytes and myofibroblast-like cells) to form contractile vessels [50,51].

Vasculogenesis: Vasculogenesis refers to the process that begins with the angioblasts: these cells, considered as endothelial cells, first form a primitive vascular network (2). The process of vasculogenesis is divided into three steps: (1) induction of haemangioblasts and angioblasts through fibroblast growth factor; (2) assembly of primordial vessels mediated by vascular endothelial growth factor; and (3) activation of corresponding receptors (FGF and VEGF) to induce the transition from vasculogenesis to angiogenesis (4).

In placenta, *de novo* formation of blood vessels starts with mesodermally derived precursor cells. Placental vasculogenesis takes place during the development of first villous vessels approximately from day 18 to day 35 *p.c.* and during the formation of mesenchymal villi from immature intermediate villi. During vasculogenesis, formation of the earliest primitive capillaries is achieved by *in situ* differentiation of hemangiogenic stem cells that are derived from pluripotent mesenchymal cells. Hemangiogenic stem cells further differentiate to hemangioblastic stem cells that give rise to angioblast cells (the progenitors of endothelial cells) and to hemangioblastic cells (the progenitors of hematopoietic cells) [52,53]. The earliest endothelial tubes are formed between day 21 and day 32 *p.c.* [14]. Fetal vascularization of the human placenta is the result of local *de novo* formation of

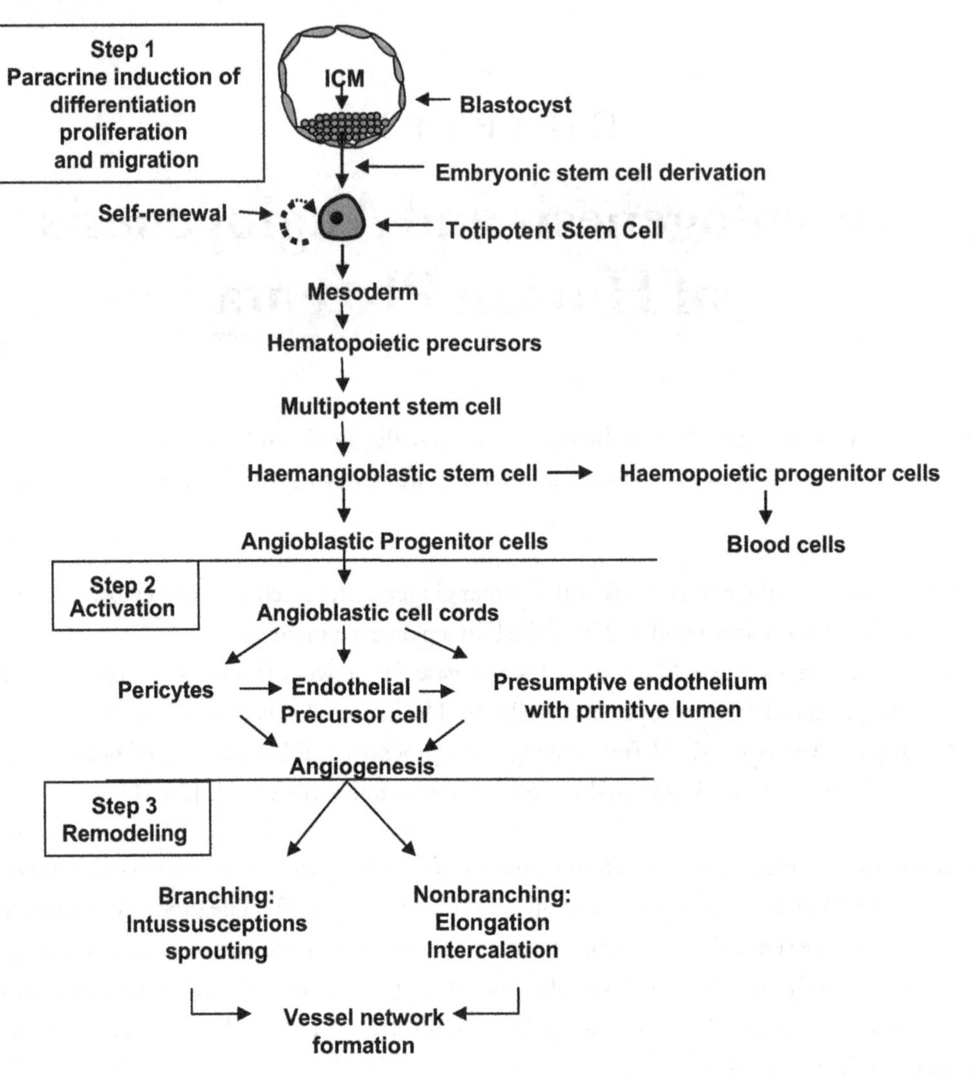

FIGURE 6.1: Proposed three steps of vasculogenesis and angiogenesis of placental villous development. The figure is modified based on references 50 and 51. ICM: inner cell mass.

capillaries from mesenchymal precursor cells in the placental villi, rather than protrusion of embryonic vessels into the placenta [50].

Angiogenesis is a physiological process involving the growth of new blood vessels from preexisting vessels. Placental vascular growth begins as early as 21 days *p.c.* and continues throughout gestation. When the primary villi are formed, the cytotrophoblast core is covered by a thick layer of syncytiotrophoblasts. With the development of secondary villi, a core of irregularly dispersed, homogenous connective tissue cells can be recognized beneath the cytotrophoblast layer. At 6 weeks of gestation, a basal lamina can be detected around the villous vessels. The development of villous capillaries from hemangioblastic cells can be observed until 10–12 weeks of gestation. From 12 weeks onwards, the capillaries coil, bulge, form sinusoids, and protrude towards the trophoblastic layers, forming a so-called syncytiocapillary membrane. Although reliable signs of formation of new vessels are absent in the second half of pregnancy, some capillary sprouts can be seen.

Around 32 days *p.c.*, villous endothelial tubes start to contact each other and the fetal allantoic vessels in the presumptive umbilical cord. A primitive fetoplacental circulation is established. From this day onward, vasculogenic *de novo* formation of capillaries followed by angiogenesis of expansion of the villous vascular system takes place until term. The angiogenic processes from day 32 until term can be considered as three periods with overlap [14]:

1. Formation of capillary networks from day 32 to 25 weeks *p.c.* by prevalence of branching angiogenesis.

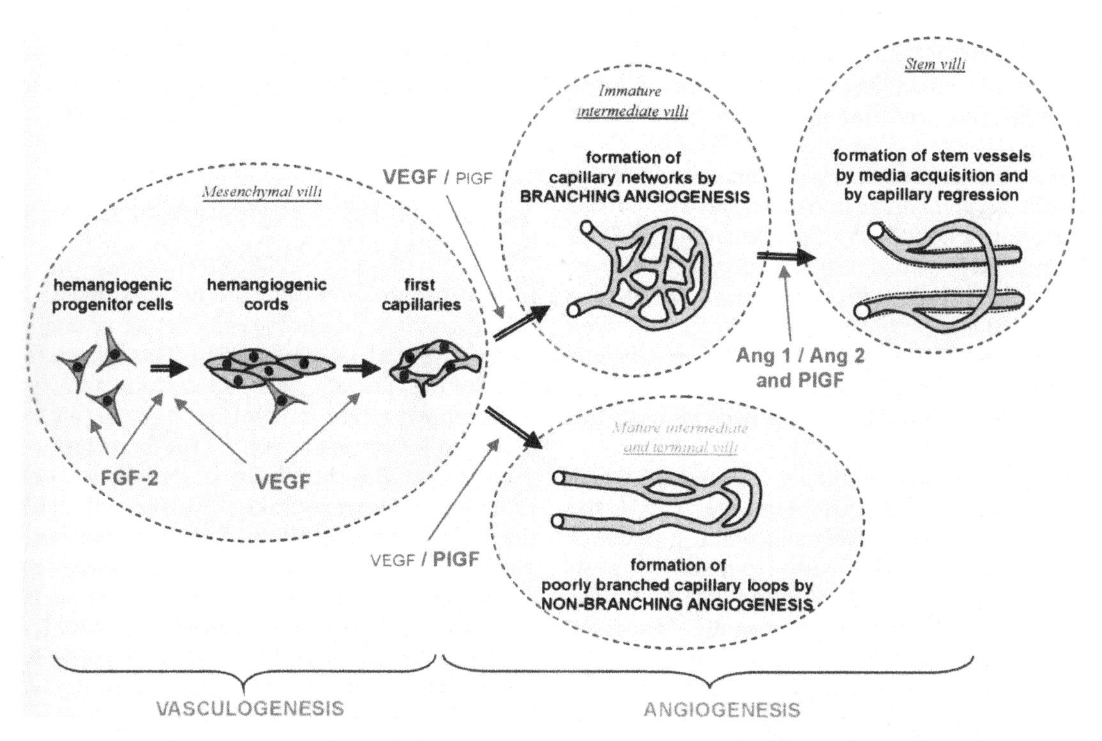

FIGURE 6.2: Proposed mechanisms of FGF, VEGF, PLGF, and angiopoietin in vasculogenesis and angiogenesis, and their distribution in villous development and presumed paracine control (red). (This figure is adopted from Benirschke K et al. Pathology of the Human Placenta 5th Edition, Figure 7.21. Reproduced from *Pathology of the Human Placenta* 5e, Benirschke et al. with permission of Springer.

2. Regression of peripheral capillary webs and formation of central stem vessels between weeks of 15 to 32 *p.c.*
3. Formation of terminal capillary loops by prevalence of nonbranching angiogenesis.

Placental vascular network development is a tightly controlled vasculogenic and angiogenic process throughout gestation. During the course of vasculogenesis and angiogenesis, angiogenic factors produced by placental cells (trophoblasts, Hofbauer cells, pericytes, and endothelial cells) play a key role. Figure 6.2 illustrates functions of VEGF, FGF, and angiopoietin associated with vasculogenesis and angiogenesis during placental vascular development.

• • • •

CHAPTER 7
Angiogenic Factors

A broad spectrum of locally produced angiogenic factors has been identified in human placenta. The most potent angiogenic factors to promote vasculogenesis and angiogenesis in the placenta include VEGF family molecules, FGF family molecules, angiopoietin/Tie system, and many others.

7.1 VEGF FAMILY MOLECUES AND VEGF RECEPTORS

The VEGF family molecules include VEGF-A, VEGF-B [54], VEGF-C, VEGF-D, and placental growth factor (PlGF). The VEGF family of receptors consists of three protein-tyrosine kinases: VEGF receptor-1 (VEGFR-1, Flt-1), VEGF receptor-2 (VEGFR-2, Flk-1/KDR), VEGFR-3, and two non-protein kinase co-receptors: neuropilin-1 and neuropilin-2, respectively. VEGFs and receptors are critical molecules that participate in vasculogenesis and angiogenesis during placental development. Interaction between VEGFR-1 and VEGFR-2 or VEGFR-2 and VEGFR-3 alters receptor tyrosine phosphorylation. Figure 7.1 illustrates the VEGF family molecules and their receptor interactions. All these VEGF and their receptors are present in the human placenta. Endothelial cells are major target for VEGF. VEGF acts as a survival factor for newly formed capillaries and has antiapoptotic effect on vascular endothelial integrity [55].

VEGF-A: VEGF-A is probably one of the most studied molecules in the VEGF family. The VEGF-A gene produces five VEGF isoforms via alternative exon splicing processes. In humans, they are VEGF121, VEGF145, VEGF165, VEGF189 and VEGF206, respectively. The differential function of these VEGF-As depends on their interaction with heparin binding to extracellular matrix and their receptors, Flt-1, Flk-1/KDR, neuropilin-1 and -2. The different solubility of each isoform also depends on the presence or absence of heparin-binding domains. VEGF-A is strongly expressed in the placental trophoblasts and villous core fetal vessel endothelium. Demir et al. examined VEGF and its receptor Flt-1 and Flk-1/KDR, as well as CD14, CD34, and CD68 expressions by transmission electron microscopy and immunohistochemistry in placental tissues from days 22–48 *p.c.* [30]. They found that VEGF is strongly expressed in villous cytotrophoblast cells and subsequently in Hofbauer cells, while its receptors Flt-1 and Flk-1/KDR are found on vasculogenic and angiogenic precursor cells. They concluded that the sequential expression of growth factors in

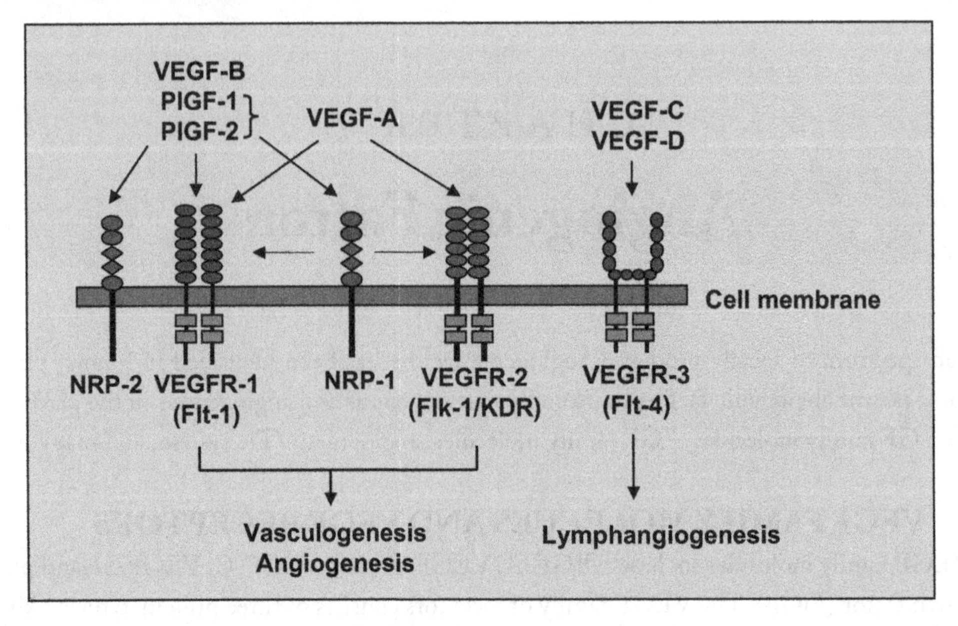

FIGURE 7.1: Schematic drawing of VEGF family molecules and their receptor interactions and signaling. VEGF-A binds to both VEGFR-1 and VEGFR-2. VEGF-B, PlGF-1 and -2 bind to VEGFR-1. VEGF-C and VEGF-D bind to VEGFR3. PlGF-1 and -2 also bind to neuropilin 1 and 2 (NRP-1 and NRP-2). Activation of VEGFR-1, VEGFR-2, and VEGFR-3 leads to autophosphorylation and coupling to the intracellular signal transducer and induce vasculogenesis, angiogenesis, and lymphangiogenesis, respectively. (The figure is modified based on Hicklin DJ and Ellis LM. *J Clin Oncol* 2005, 23:1011–1027.) Published with permission of the American Society of Clinical Oncology.

different cell types point out the fact that placental vasculogenesis and angiogenesis are clearly distinct events (30). Their data also showed trophoblasts express and produce soluble VEGF isoforms, which imply that the developmental expression and secretion of VEGF involve recruitment, maintenance, and formation of earlier angiogenic cells and vessels. Interactions between VEGF and Flk-1/KDR and Flt-1 regulate placental vasculogenesis and angiogenesis in paracrine and/or autocrine manners. Figure 7.2 shows CD31 staining in villous tissues from first-, second-, and third-trimester placentas. The localization of fetal vessels close to the trophoblast layer in the first-trimester placenta indicates the important role of trophoblasts in vessel formation.

VEGF-B: Several studies have found positive VEGF-B expression in the human placenta [54,56,57]; however, its specific function in placenta development has not been defined.

VEGF-C: VEGF-C was purified and cloned from the human prostatic carcinoma cell cDNA [58]. VEGF-C mediates many functions besides angiogenesis in the placenta [59]. hVEGF-C has a

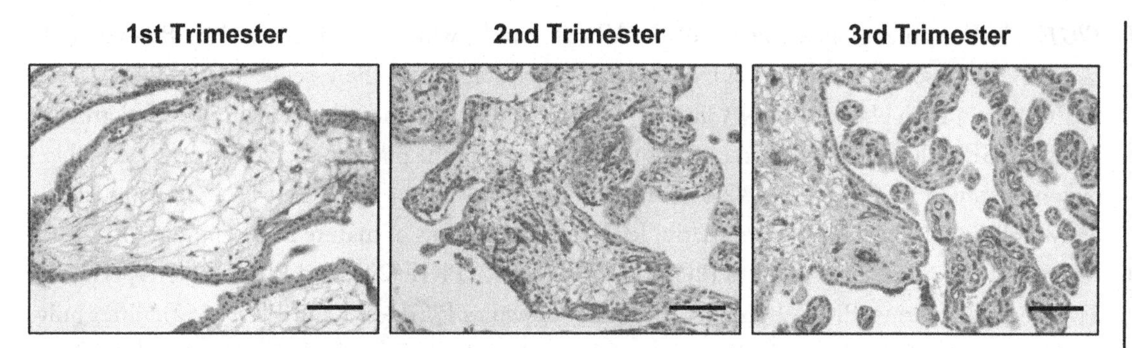

FIGURE 7.2: CD31 staining in villous tissue sections from first-, second-, and third-trimester placentas. Bar = 20 micron.

molecular weight of 23kDa, and its amino acid sequence is 31%, identical to hVEGF-A165 [60], 27% to VEGF-B167 [54], and 25% to PlGF-1 [61]. An *in vitro* cytotrophoblast study has shown that removal of VEGF-C, PlGF, and/or angiopoetin-2 (Ang2) from culture medium caused a significant increase in cytotrophoblast apoptosis [55]. A recent study also shows that VEGF-C could facilitate immune tolerance of human uterine NK cells at the maternal-fetal interface [59]. The cell protective effects of VEGF-C are attributed by the induction of antigen peptide transporter 1 (TAP-1) expression and MHC class I assembly in target cells [59]. Hypoxia is an important regulator of VEGF-C expression and production in the placenta [62,63].

VEGF-D: VEGF-D is cloned from human lung cDNA [64], with 48% identity with hVEGF-C, 31% identity with hVEGF-A165, 28% identity with hVEGF-B167, and 32% identity with hPlGF [64]. Both VEGF-C and VEGF-D are rich in cysteine residues in their C-terminal regions [64]. Recently, a mouse protein designated the c-Fos-induced growth factor was reported that had 85% amino acid identity with hVEGF-D [65]. It is highly likely that c-Fos-induced growth factor is a mouse homologue of VEGF-D. VEGF-D stimulates growth of vascular and lymphatic endothelial cells, but VEGF-D is dispensable for development of the lymphatic system [66]. Both VEGF-C and VEGF-D are potent mitogens to endothelium and ligands for VEGFR-2 (KDR) and VEGFR-3 (Flt-4) [64]. Human placenta expresses VEGF-C, VEGF-D, and VEGFR-3. VEGF-C is mainly expressed by villous core fetal vessel endothelium [67], whereas VEGF-D and VEGFR-3 are present in both syncytiotrophoblasts and villous core fetal vessel endothelium [67]. The differential expression between VEGF-C and VEGF-D suggests the potential disparity of the two VEGF isoforms for trophoblast function and fetal vessel development. Although the placenta has no known lymphatic circulation, the expression of lymphatic markers in the placental trophoblasts, villous stromal, and villous endothelium imply that lymphatic structure exists in the placenta. (See later placenta lymphatic section.)

PlGF: PlGF is an angiogenic factor of the VEGF family, which was discovered by Persico et al. in 1991. They isolated a human cDNA coding protein related to the vascular permeability factor (VPF) from a term placenta cDNA library [61] and therefore named this protein placental growth factor (PlGF). PlGF is highly homologous (53% identity) to the platelet-derived growth factor-like region of human VEGF and has 42% amino acid homology with VEGF-A [61]. In humans, alternative splicing of PlGF primary transcript leads to four isoforms of mature PlGF proteins, PlGF-1, -2, -3, and -4, which correspond to PlGF-131, PlGF-152, PlGF-203, and PlGF-224, respectively. Two dominant forms, PlGF-1 and PlGF-2 (also known as PlGF-131 and PlGF-152), differ only by the insertion of a highly basic 21-amino acid stretch at the carboxyl end of the protein [68]. This additional basic region contains a heparin-binding motif and allows PlGF-152 to bind to heparin. Both PlGF-1 and PlGF-2 bind to VEGFR-1, whereas PlGF-2 is also able to bind to neuropilin-1 and neuropilin-2, and to heparin sulfate proteoglycans (HSPG) [69] due to heparin-binding motif in its carboxyl terminal sequence.

PlGF functions to control trophoblast growth and differentiation [70,71]. Immunostaining studies showed PlGF is localized on trophoblasts and villous core endothelial cells [56]. In mice, abundant transcripts encoding PlGF were found in trophoblast giant cells associated with the parietal yolk sac at early stages of embryogenesis [72]. The secretion of PlGF by trophoblast giant cells may well be the signal that initiates and coordinates vascularization in the decidual and placenta during early embryogenesis [72]. An animal study has shown that $PlGF^{-/-}$ mice are born without apparent vascular defects. However, loss of PlGF impairs angiogenesis, plasma extravasation, and collateral growth during ischemia, inflammation, and wound healing. Transplantation of wild-type bone marrow rescued the impaired angiogenesis and collateral growth in $PlGF^{-/-}$ mice, indicating that PlGF might have contributed to vessel growth in the adult by mobilizing bone-marrow-derived cells [73].

PlGF has been proposed to stimulate angiogenesis by displacing VEGF from the "VEGFR-1 sink," thereby increasing the fraction of VEGF available to activate VEGFR-2. Alternatively, PlGF might stimulate angiogenesis by transmitting intracellular signals through VEGFR-1. Synergism between VEGF and PlGF contributes to both physiological angiogenic processes and pathological plasma extravasation [73]. The synergism between PlGF and VEGF is specific, as PlGF deficiency impairs cell response to VEGF. In $PlGF^{-/-}$ cells, VEGF only induced endothelial migration by 25%, proliferation by 40% and survival by 7% compared to $PlGF^{+/+}$ endothelial cells [73]. Furthermore, high concentrations of PlGF alone did not affect the response of wild-type endothelial cells to VEGF, presumably because these cells already produced sufficient PlGF. However, at low doses, PlGF dose-dependently restored the impaired VEGF-response in $PlGF^{-/-}$ endothelial cells. Antibodies specific for VEGFR-1 completely block the response of $PlGF^{-/-}$ cells to VEGF and PlGF, indicating that PlGF activates VEGFR-1. VEGF/PlGF heterodimers also stimulates the survival of $PlGF^{-/-}$ in endothelial cells. Thus, PlGF, though ineffective alone, amplifies the VEGF response by activating VEGFR-1 [73].

There are probably several mechanisms that are involved in angiogenic function of PlGF: (1) PlGF binds to VEGFR-1 in endothelial cells which facilitates VEGF binding and activation of VEGFR-2; (2) recruitment of monocytes/macrophages to promote vessel growth; and (3) by mobilizing hematopoietic progenitor cells from bone marrow [69]. PlGF and VEGF-A are also known to form a PlGF/VEGF heterodimer, which is a highly potent endothelial mitogen [73].

Altered VEGF/PlGF/sFlt-1 expression and production: As discussed earlier, physiological hypoxia or low oxygen tension plays a critical role in early placental development. However, persistent hypoxic environment as a result of improper remodeling of the decidual spiral arterioles will lead to imbalanced angiogenic process, especially during the second half of gestation in human pregnancy. Hypoxia promotes VEGF and VEGFR-1 (Flt-1), but down-regulates PlGF, expression and production [62,74]. Altered Flt-1 expression not only affects villous vessel development, but also results in release of an excess amount of its soluble form (sFlt-1) into the maternal circulation. Increased sFlt-1 production by trophoblasts is believed to be one of the critical pathogenic factors in the development of preeclampsia. Several groups including our own have reported that sFlt-1 gene products are up-regulated in preeclamptic placentas [62,74-76]. The excess sFlt-1 noted in the bloodstream is associated with decreased free VEGF and PlGF [77,78]. sFlt-1 neutralizes the effects mediated by VEGF and PlGF in systemic vasculature. Placental-derived sFlt-1 could pose detrimental effects on maternal vasculature, because of its strong antiangiogenic activity.

7.2 FGFS AND FGF RECEPTORS

Fibroblast growth factor (FGF) was originally found in pituitary extracts by Armelin in 1973 [79], who showed that pituitary extracts stimulate 3T3 cell growth, an established mouse fibroblast cell line [79]. Gospodarowicz and colleagues also found that cow brain extract causes fibroblasts to proliferate [80,81]. To date, 22 members of the FGF family and 4 members of FGF receptors have been identified in humans. The FGF family molecules have molecular weights of approximately 14–16kDa [80,81]. Based on their structural, functional and characteristic differences, they are divided into 3 groups. FGF1–10 bind to FGF receptors. FGF1 is known as the "acidic" form and FGF2 is known as the "basic" form. FGF11–14 are known as FGF homologous factors 1–4 (FHF1-4). They do not bind to FGF receptors, but rather to islet brain-2 (IB2) and their function is involved in intracellular kinase scaffold protein regulation [82]. FGF16–23 are not yet well characterized. Humans have no FGF15. Among these FGFs, FGF1 and FGF2 are the best studied. Figure 7.3 shows the canonical FGF-FGFR binding at cell membrane. One of the major functions of FGF1 and FGF2 is their potent angiogenic potential. They promote endothelial cell proliferation

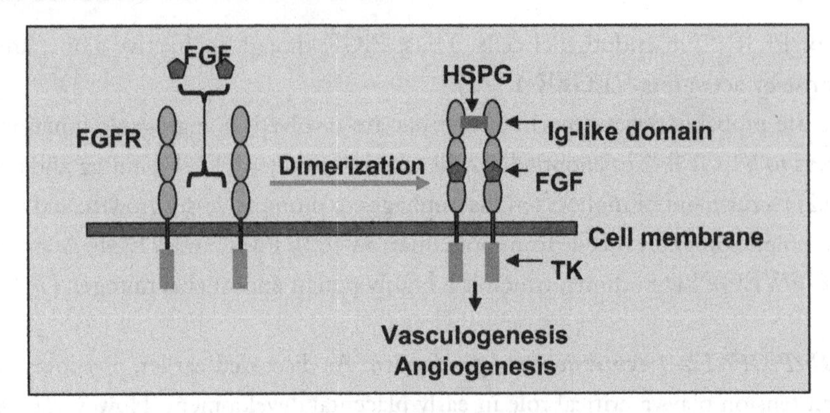

FIGURE 7.3: Schematic drawing of canonical FGF-FGFR signaling. The extracellular domain of FGFR consists of three immunological-like (Ig) domains, designated as D1, D2, and D3. In canonical FGF signaling, FGF binds to D3. The D3 domain has splicing variants, which provide specific recognition for different FGFs. Binding of FGF to FGFR results in receptor dimerization and formation of FGF/FGFR/HSPG complex at the cell surface, which induce transphophorylation of the intracellular tyrosine kinases, followed by the activation of cytosolic signaling events. Heparan sulfate proteoglycan (HSPG) is a coreceptor of FGFR.

and vasculogenesis. An animal study has shown that FGF-1 and FGF-2 (bFGF) are more potent angiogenic factors than VEGF-A or platelet derived growth factor (PDGF) [83].

FGFs are heparin-binding proteins and interact with cell-surface associated heparan sulfate proteoglycans as essential for FGF signal transduction. FGFs are key players in the processes of proliferation and differentiation of widely diverse cells and tissues. FGFs are multifunctional proteins stimulating a broad variety of biological effects including pluripotency and angiogenesis. The placenta is a rich source of FGFs [84]. FGFs are essential for trophoblast stem cell function, as studies have shown that trophoblast stem cells require FGF4 for self-renewal and prevention of terminal differentiation. FGF-10 stimulates trophoblast migration/invasion by eightfold in an *in vitro* villous explant outgrowth study [34]. Shams and Ahmed examined FGF expression in the first trimester and term human placentas by *in situ* hybridization [85]. They found that bFGF expression was seen in syncytiotrophoblasts surrounding the placental villi, and in cytotrophoblast cells of first-trimester placentas. At term, the bFGF gene expression is detected in syncytiotrophoblasts and in fetal membranes. Strong expression of bFGF mRNA is also detected in the smooth muscle cells around the mid- and large-sized placental vessels [85]. Colocalization of acidic FGF and basic FGF is also observed [86]. Anteby et al. analyzed the expressions of FGF-10 and FGFR1–4 in the first-, second- and third-trimester placentas, as well as in the decidua [87]. They found that FGF-10 was expressed in decidual cells and in cytotrophoblasts in placentas throughout gestation, while FGFR1–4 were expressed in placenta but not decidua [87]. Placental expression of FGFRs was temporally regulated. In the first-trimester placenta, FGFR 1–4 are expressed by Hofbauer cells, while FGFR-1 and FGFR-4 are expressed in cytotrophoblast columns and syncytiotrophoblasts. Similar expression is seen in second-trimester placentas with additional expression of FGFR-1 in the blood vessel walls. The expressions of FGFR-1 and FGFR-4 in the third trimester are comparable to that seen in the second trimester. The abundance of FGFR expression in Hofbauer cells implies that mesenchymal trophoblast interactions are important for regulation of villous development [87].

7.3 ANGIOPOIETINS AND TIE SYSTEM

The angiopoietin signaling system includes four ligands (angiopoietin-1, angiopoietin-2 and angiopoietin-3/4) and two corresponding tyrosine kinase receptors, Tie-1 and Tie-2. Using a novel expression cloning technique in intracellular trapping and ligand protection, Davis et al. identified angiopoietin-1 (Ang-1) as a secreted ligand for Tie-2 [88]. The human gene for Ang-1 encodes a 498-amino acid polypeptide with predicted coiled-coil and fibrinogen-like domains with a molecular weight of 57kDa [88]. The protein encoded by this gene is a secreted glycoprotein that activates the receptor by inducing its tyrosine phosphorylation. Angiopoietin-2 (Ang-2) was identified by homology screening and was shown to be a naturally occurring antagonist for Ang-1 at the Tie-2 receptor [89]. The Ang-2 protein is 496 amino acids in length with a secretion signal peptide [89]. Human and mouse Ang-2 are 85% homologous to amino acid sequence and about 60% identity with Ang1 [89].

The angiopoietin-Tie system is a vascular specific ligand/receptor system, which controls endothelial cell survival and vascular maturation and plays an important role in the development of blood and lymphatic vasculature. Ang-1 is chemotactic for endothelial cells and promotes angiogenesis. Both Ang-1 and Ang-2 bind to Tie-2. The binding process plays a critical role in mediating reciprocal interactions between the endothelium and surrounding matrix and mesenchymal tissues. The protein also contributes to blood vessel maturation and stability. The human placenta expresses both Ang-1 and Ang-2 [89]. Ang-1 is present in the syncytiotrophoblasts, whereas Ang-2 is expressed in both cytotrophoblasts and syncytiotrophoblasts [90]. Tie-1 and Tie-2 expressions are also found in different compartments in the placenta. Tie-1 is primarily expressed in the syncytiotrophoblasts, while both cytotrophoblasts and syncytiotrophoblasts express Tie-2. In addition, villous core fetal vessels express both Tie-1 and Tie-2 [90]. Placental Ang-1 and Ang-2 expressions were reduced with increasing gestational age [91]. Ang-1 and Ang-2 activate trophoblast Tie-2 to promote growth and migration during placental development. An *in vitro* study has shown that Ang-1 stimulated trophoblast migration and Ang-2 triggered the release of NO [92].

Tie-2-dependent activation of RhoA and Rac1 participates in endothelial cell motility, a process triggered by angiopoietin-1 [93].

The angiopoietin/Tie system is important in vessel maturation and plays a role in regulation of vascular smooth muscle cell recruitment. This system is also involved in maintaining blood vessel quiescence. The function of Ang-1 and Ang-2 depends on the receptor types. Binding of Ang-1 or Ang-2 to Tie-2 induces Tie-2 dimerization and phosphorylation, subsequently activates signaling pathways to initiate angiogenesis, while binding of Ang-2 to Tie-1/Tie-2 heterodimer could not induce receptor phosphorylation, Tie-2 activation will not occur [94] (Figure 7.4).

Studies also showed that the functional consequences of angiopoietin/Tie system are integral to VEGF-A responsiveness. If VEGF-A is present, Ang-2 will enable endothelial cell migration and proliferation, and therefore angiogenesis. If VEGF-A is inhibited, Ang-2 will lead to endothelial cell death and vessel regression [95,96]. The spatiotemporal regulation of receptors and ligands, as well as their integral interaction and regulation with VEGF and FGF systems, provide a unique role of the angiopoietin/Tie system in the placental vascular development.

Two molecular mechanisms for Ang-2 -mediated blood vessel remodeling have been proposed, Tie-2-dependent and Tie-2-independent mechanisms. In the Tie-2-dependent pathway, Ang-2 antagonizes Ang-1-mediated phosphorylation of Tie-2 receptor [97]. Ang-1 is responsible

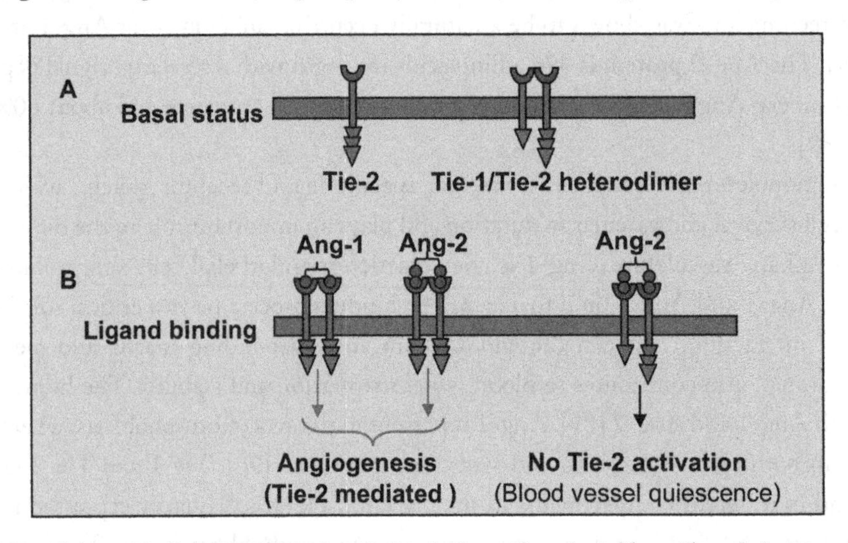

FIGURE 7.4: Schematic drawing of angiopoietin-1 (Ang-1) and -2 (Ang-2) and their receptor Tie-1 and Tie-2 signaling. A: Under basal condition, Tie-2 is located on the cell membrane either as a monomer or form heterodimer with Tie-1. B: Binding of Ang-1 or Ang-2 to Tie-2 induces Tie-2 dimerization and phosphorylation, and subsequently activates signaling pathways to initiate angiogenesis, including vessel sprouting. However, binding of Ang-2 to Tie-1/Tie-2 heterodimer could not lead to the receptor activation. The blood vessel quiescence is maintained.

for blood vessel stabilization and maturation by recruitment of pericytes and smooth muscle cells [98]. The vasculopathy phenotype in Tie-2-deficient [99] and Ang-1-deficient mice is exacerbated in Ang-2 transgenic mice [97]. This vasculopathy is characterized by endothelial cell detachment and disrupted blood vessel formation [97]. Manipulation of Ang-2 levels by direct placental transfection induces a similar phenotype in a murine model [100]. The Tie-2-independent mechanism is suggested based on Ang-2 modulation of endothelial adhesion. It was found that Ang-2 binds to $\alpha V\beta 5$ and vitronectin integrins, which mediates migration and spreading of both endothelial and nonendothelial cells [101]. An animal study has also shown that Ang-2 gene delivery to a pregnant uterus results in modulation of fetal and maternal placental vessels coupled with $\alpha V\beta 5$ and vitronectin expressions in placental trophoblasts [100]. Furthermore, site-specific placental gene transfer of Ang-2 results in an increase in both placental weight and fetal weight, indicating that Ang-2 plays a significant role in placental angiogenesis and vascular development [102].

• • • •

CHAPTER 8

Vasoactivators and Placental Vasoactivity

Fetal homeostasis depends on the efficiency of the maternal-fetal circulation. As discussed early, maternal arterial pressure propels maternal blood flow through the placenta. The low-resistance uteroplacental vessels accommodate the increased perfusion need for the development of placental vasculature throughout the course of gestation. Therefore, uteroplacental and villous core vasoactivities are essential in maintaining the high-flow maternal blood perfusion into the low-resistance placental intervillous space. Placental vascular reactivity is controlled by several vasodilator and vasoconstrictor systems including the renin-angiotensin system, arachidonic metabolites (thromboxane and prostacyclin), endothelin and its receptors, and NO. Thus, the balance between vasodilators and vasoconstrictors from both the maternal and placental compartments is critical for the homeostatic balance of placental vascular function. Clearly, increased vasoconstriction of the placental vasculature leads to an abnormal course of pregnancy, such as preeclampsia and IUGR. Placental vessels lack autonomic innervation. Hence, humoral effects from the placenta or vasoactivators produced by the placental cells dominate the regulation of fetoplacental vascular reactivity, which is mediated by autocrine and paracrine regulatory mechanisms. Numerous vasoactivators and their corresponding receptors are present in the placenta, and differential synthesis and actions of known vasoactivators are found along the umbilical cord, chorionic plate, and villous vessels (Figure 8.1) [103].

8.1 RENIN-ANGIOTENSIN SYSTEM

The renin-angiotensin system (RAS) or the renin-angiotensin-aldosterone system (RAAS) is an autacoid system that regulates blood pressure, sodium, and fluid homeostasis. When blood volume or Na^+ concentration is low, the kidneys secrete renin. Renin cleaves the precursor molecule angiotensinogen, a serum α2-globulin produced in the liver, and converts it to angiotensin I. Human angiotensinogen is 452 amino acids long, and the first 12 amino acids are critically important for its activity. Angiotensin I is 10 amino acids with low biological activity and exists mainly as a precursor to angiotensin II. Angiotensin-converting enzyme (ACE) converts angiotensin I to angiotensin II. ACE is present on all endothelial cells but with highest density in the lung capillaries. Angiotensin II is the major bioactive product of the renin-angiotensin system. In the kidney, angiotensin II has a direct effect on the proximal tubules to increase Na^+ reabsorption. By binding to its receptors on

	Vasoconstrictors	Vasodilators
Umbilical vessels	Ang II ET-1(ETRA) TXA_2 BK 5-HT AVP	PGI_2 NO AM (ET-1/ETR_2)
Chorionic plate vessels	Ang II ET-1 TXA_2 5-HT	NO PGI_2 AM
Villous tree	Ang II ET-1 TXA_2 BK 5-HT LTs ROS	NO ANP PGI_2 AM

FIGURE 8.1: Different actions of vasoconstrictors and vasodilators on umbilical cord and placental vessels. This figure is modified based on reference 102. Ang II: angiotensin II; ET-1: endothelin-1; TXA_2: thromboxane A_2; BK: bradykinin; 5-HT: 5-hydroxytryptamine; AVP: arginine vasopressin; PGI_2: prostacyclin; NO: nitric oxide; AM: adrenomedullin; LTs: leukotriene(s); ROS: reactive oxygen species; ANP: atrial natriuretic peptide.

mesangial cells, angiotensin II causes these cells to contract along with the blood vessels surrounding them, and promotes the release of aldosterone from the zona glomerulosa in the adrenal cortex to reduce glomerular filtration and renal blood flow and fluid resorption. Angiotensin II is a potent vasoconstrictor that constricts both arteries and veins to increase blood pressure by binding to its receptor AT-1. Angiotensin II has a very short half-life in the circulation, ~30 seconds, while in tissues, its half-life is about 15–30 minutes. Besides angiotensin II {(Ang-(1–8)}, other angiotensin peptides, such as angiotensin III {Ang-(2–8)}, angiotensin IV {Ang-(3–8)}, and angiotensin-(1–7) also have biological activities. For example, angiotensin-(1–7) has become an angiotensin isoform of interest in the past few years, because its cardiovascular and baroreflex actions counteract those of angiotensin II [104]. The metabolic pathway of RAS is shown in Figure 8.2.

During normal pregnancy, maternal RAS components including angiotensin I, angiotensin II and angiotensin 1–7 levels, as well as angiotensinogen and renin activities are all increased compared to the nonpregnant state [105]. Women with uncomplicated pregnancy are normotensive despite a twofold increase in angiotensin II levels and increased circulating blood volume, suggesting that in

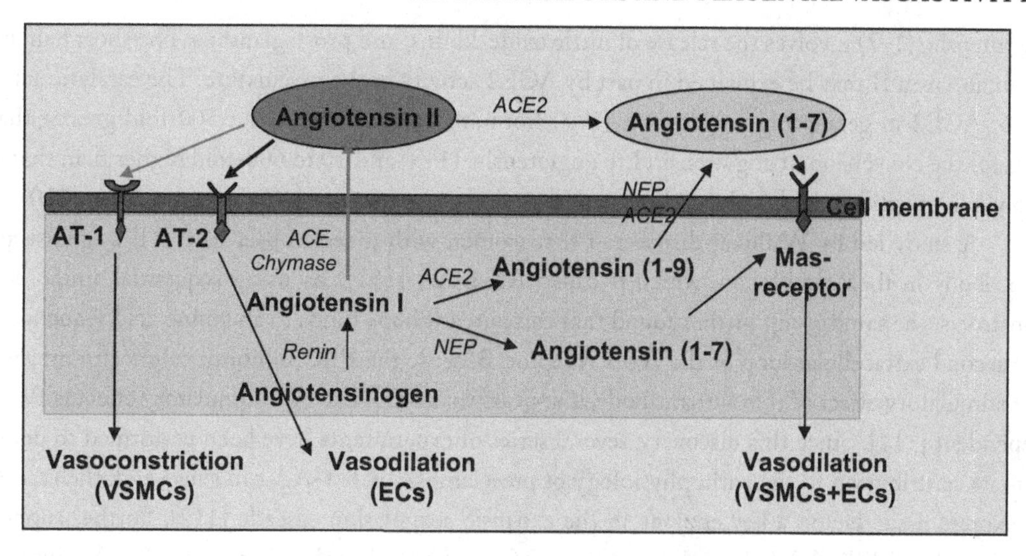

FIGURE 8.2: Renin-angiotensin system (RAS) pathway. Renin is produced from prorenin through proteolytic removal of prosegment. The active renin converts angiotensinogen to angiotensin I. Angiotensin I is then converted either to angiotensin II by angiotensin II converting enzyme (ACE) and chymase or to angiotensin (1–9) by ACE2. Angiotensin II binding to AT-1 on vascular smooth muscle cells induces vasoconstriction, and binding to AT-2 on endothelial cells induces vasodilation. ACE, ACE2, and NEP work together to produce angiotensin (1–7) directly or indirectly from both angiotensin I and angiotensin II. Binding of angiotensin (1–7) to Mas receptor induces vasodilation.

normal pregnancy the pressor effects of angiotensin II are somewhat compensated [106–108]. Although increased RAS signals with less pressor effect in normal pregnancy is paradoxical, elevated progesterone levels, increased ACE2, and angiotensin 1–7 activities may be the mechanisms for explaining the relative refractoriness to, and less pressor effects by angiotensin II stimulation. Interestingly, normal pregnant women lose pregnancy-acquired vascular refractoriness to angiotensin II within 15–30 minutes after the placenta is delivered [108]. This observation indicates a rapid clearness of substances originating in the placenta. Therefore, placenta-derived progesterone seems a likely candidate for this role [108]. Increased ACE2 and angiotensin 1–7 activities could also be major components counteracting angiotensin II during pregnancy. ACE2 is a carboxypeptidase with 42% homology with ACE, but it exerts different biological activities from ACE. ACE2 generates angiotensin-(1–7) from both angiotensin II and angiotensin I. ACE2 cleaves one amino acid from angiotensin II to form angiotensin-(1–7). ACE2 could also cleave one amino acid from angiotensin I to generate angiotensin-(1–9), which can further be converted to angiotensin-(1–7) by neprilysin and ACE. ACE2 exhibits a high catalytic efficiency to generate angiotensin-(1–7) and at the same time inactivates the vasoconstrictor counterpart angiotensin II [109]. The vasodilatory activity of angiotensin-(1–7) involves the release of nitric oxide, kinins, and prostaglandins. The short half-life of angiotensin II may be explained in part by ACE2 activity in the vasculature. The catalytic activity of ACE2 in generating angiotensin-(1–7) from angiotensin II is about 500-fold greater than that for the conversion of angiotensin I to angiotensin-(1–9) and 10 to 600-fold higher than that of prolyl oligopeptidase and prolyl carboxypeptidase to form angiotensin-(1–7), respectively [110].

A study led by Wallukat discovered that women with preeclampsia have AT-1 agonist autoantibody in their circulation, which is called AT-1-AA [111]. By using sequential amino acid constructs, the same group further found that this autoantibody binds to an amino acid sequence of the second extracellular loop of the AT-1 receptor. Because the PKC inhibitor calphostin prevents the stimulatory effect of this autoantibody, it appears that the AT-1-AA stimulatory effect is PKC-dependent [111]. Since this discovery, several series of experiments have been performed to delineate its contribution to the pathophysiology of preeclampsia. AT-1-AA can cause endothelial cells to express tissue factor, a key enzyme in the extrinsic coagulation cascade [112]. Further support for the role of AT1-AA in preeclampsia comes from the studies showing that AT-1-AA induces oxidative stress including stimulation of NADPH oxidase and activating NF-κB [113]. In addition to oxidative stress and coagulation cascades, AT-1-AA also affects other cellular processes. For example, autoantibody from preeclamptic women is capable of activating angiotensin receptors on placental trophoblasts [114]. The presence of AT-1-AA provides one more potential explanation of the increased vascular reactivity seen in women with preeclampsia.

Fetal placental tissue has its own local RAS system. That is, all components of classical RAS system renin, angiotensinogen, angiotensin I, angiotensin II, angiotensin 1–7, ACE, ACE2, as well as receptors AT-1 and AT-2, have been shown to be present in the fetal placental unit. In 1967, Hodari et al. identified a renin-like substance in human placental tissue [115]. Later, renin, angiotensinogen, ACEs, and angiotensin I and II were all confirmed in the human placenta [116–118]. Kalenge et al. analyzed concentrations of active renin, prorenin, ACE, and angiotensin II in placental tissue and fetal membrane homogenates, and detected all these RAS components in both placental tissues and fetal membranes [117]. They also noticed relative higher values of renin, ACE, and angiotensin II in tissue homogenates from placentas delivered by women with pregnancies complicated with preeclampsia [117]. ACE activity and protein expression are more concentrated in the villous core vessel fraction than in trophoblasts or macrophages [119]. Herse et al. found that messenger RNA expressions for renin, ACE, and angiotensinogen are dominantly expressed in decidua tissues as compared to placental tissues from both normal and preeclamptic pregnant women. In contrast, strong AT-1 receptor expression is found in the placenta verses decidua tissue from both normal and preeclamptic pregnancies [120]. The immunoreactivities for both angiotensin-(1–7) and ACE2 are localized to syncytiotrophoblasts, cytotrophoblasts, villous core fetal vessel endothelium, and vascular smooth muscle cells. Angiotensin-(1–7) expression is reduced in placentas from preeclampsia. We examined AT-1 and AT-2 expressions in first-, second-, and third-trimester placentas and found that in the first-trimester placentas, AT-1 and AT-2 receptors were expressed to both the cyto- and syncytiotrophoblasts. In the second trimester placentas, AT-2 expression was more pronounced in villous stroma and AT-1 was on the apical membrane of syncytiotrophoblasts. In the full-term placentas, strong immunolabeling of these two receptors was observed on villous core fetal vessels. Syncytiotrophoblasts expressed AT-1, but not AT-2 (Figure 8.3). The intense AT-1 and AT-2 immunostaining seen in villous core fetal vessel endothelium in third-trimester placentas suggest the role of angiotensin II receptors in relaxation of fetal vessels since activation of AT-1 and AT-2 receptors on endothelial cells results in production of vasodilatory agents, nitric oxide, and prostacyclin (PGI_2), which counteract the direct vasoconstrictor effects of angiotensin II on the adjacent smooth muscle cells.

In preeclampsia, the local placental RAS differs from that in the circulation. By measuring the renin protein and its activity in the supernatants of placental fresh homogenates, Singh et al. found that preeclamptic placentas produced more renin and had elevated renin activity

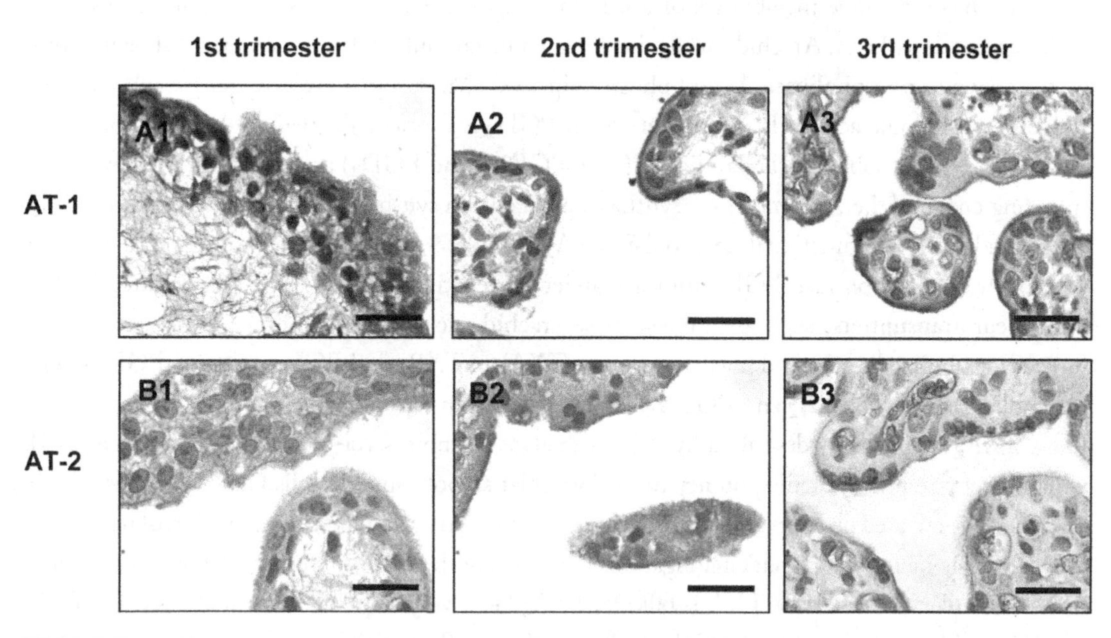

FIGURE 8.3: Immunostaining of angiotensin II receptors AT-1 and AT-2 in the first-, second-, and third-trimester placentas. A: AT-1; and B: AT-2. 1: First-trimester placentas; 2: Second-trimester placentas; and 3: Third-trimester placentas. Please note the transition of AT-1 and AT-2 receptor expressions in trophoblasts of the first-trimester placenta to villous core fetal vessel endothelium of the third-trimester placentas. Bar = 25 micron.

[121]. Higher angiotensin II levels and increased ratio of angiotensin I to angiotensin-(1–7) expression were also found in placental tissues from preeclampsia [122,123]. The presence of AT-1 autoantibody in the preeclamptic placenta further contributes to the altered RAS system in preeclamptic placentas [122]. Therefore, altered placenta RAS system is considered to attribute to the increased vasoconstriction and placental pathophysiology in preeclampsia. Using a unique organ bath perfusion model, we have shown that vasoactivity of chorionic plate arteries from preeclamptic placentas is enhanced compared to that from normal placentas, and that the preeclamptic placenta-derived factors induced chorionic plate artery contraction can be attenuated by AT-1 receptor blocker losartan [124]. These observations indicate that altered RAS pathway metabolites and increased angiotensin II generated within placental tissue contribute to increased vasoconstriction in preeclampsia.

8.2 ARACHIDONIC ACID METABOLITES: THROMBOXANE AND PROSTACYCLIN

Thromboxane (TXA_2) and prostacyclin (PGI_2) are members of the family of lipids known as eicosanoids. They are native metabolites of arachidonic acid (5,8,11,14-eicosatetraenoic acid) via the cyclooxygenase pathway. Arachidonic acid is released intracellularly from membrane phospholipids by the enzymes phospholipase A_2 and phospholipase C. Microsomal cyclooxygenase subsequently converts arachidonic acid to PGG_2, from which PGH_2 is formed. Thereafter, three end products, TXA_2, PGI_2, and stable prostaglandins (PGE_2, $PGF2\alpha$, and PGD_2) are generated by their corresponding enzymes, i.e., thromboxane synthase, prostacyclin synthase, and isomerases from PGH_2. TXA_2 and PGI_2 are functional antagonist. TXA_2 stimulates platelet aggregation and is a potent vasoconstrictor; In contrast, PGI_2 inhibits platelet aggregation and is a potent vasodilator. Unlike other neurotransmitters, such as amines, these arachidonic acid metabolites are not stored, but further metabolized to the end inactive product: TXA_2 to TXB_2 and PGI_2 to 6-keto-$PGF1\alpha$. The metabolic pathway of TXA_2 and PGI_2 synthesis is shown in Figure 8.4.

TXA_2 was initially described by Palmer et al. as a rabbit aorta-contracting substance [125], which has a potent contractile potency toward vascular smooth muscle cells [126]. TXA_2 is mainly produced by activated platelets and has prothrombotic properties, stimulating activation of new platelets, as well as increasing platelet aggregation. Platelets are the main source of TXA_2 in the circulation. The molecular weight of TXA is 60kDa. TXA_2 has a very short half-life of about 30 seconds, and it is rapidly hydrolyzed to the biologically inactive TXB_2, which is stable and often used as an indicator of TXA_2 production/level in biological samples. When TXA_2 is produced from PGH_2 by thromboxane synthase, 12-**L**-hydroxy-5,8,10-heptadecatrienoic acid (HHT) and malondialdehyde (MDA) are also simultaneously produced [127].

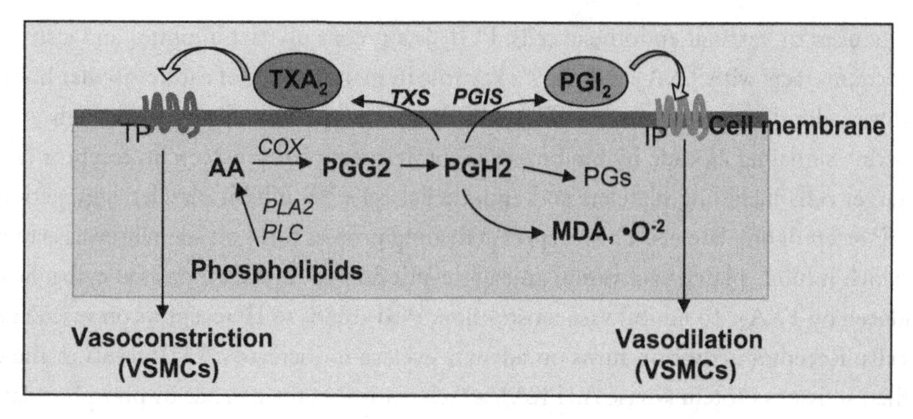

FIGURE 8.4: Arachidonic acid – TX/PGI pathway. Phospholiase A_2 (PLA_2) and phospholipase C (PLC) liberate arachidonic acid (AA) from membrane lipid of placental syncytiotrophoblasts. Cycloxygenase (COX) produces PGG_2 and PGH_2 from AA. Subsequently, thromboxane (TXA_2), prostacyclin (PGI_2), and prostaglandins (PGs) are produced via corresponding synthases, TX synthase (TXS), PGI_2 synthase (PGIS), and prostaglandin isomerases, respectively. The function of TXA_2 and PGI_2 is opposite. TXA_2 binds to its TP receptor inducing vasoconstriction and PGI_2 binds to its IP receptor causing vasodilation.

Because of its short half-life, TXA_2 primarily functions as an autocrine or paracrine mediator by binding to its receptor, thromboxane receptor (TP receptor), in the nearby tissues such as on platelets and endothelial cells surrounding its site of production. TP receptor is a G-protein coupled receptor. The gene for the TP receptor is located on chromosome 19. Two TP receptor isoforms have been found [128]. One was cloned from human placenta and the other from human endothelial cells. The former was referred as TP alpha and the latter as TP beta. Both isoforms are present in platelets [128]. They differ only in their carboxyl-terminal tails, but their function is different. TP alpha activates adenylyl cyclase, while TP beta inhibits it. Different G-protein receptor activation accounts for the downstream effects of TP-receptor responses. In fact, early work has shown that TP-mediated platelet shape changes are mainly dependent on G12/13 activation, while Gq activation is responsible for platelet aggregation [129]. The mechanism of TXA_2 elicited vascular smooth muscle constriction is involved in Ca^{2+} signaling. TP receptor activation causes phospholipase C-catalyzed phosphoinositide hydrolysis, which in turn mobilizes intracellular Ca^{2+} and Gq acts as a trimeric G protein coupled to TP receptor to activate phospholipase C [130].

Prostacyclin was discovered by Moncada et al. in 1976 [131]. They identified a lipid mediator that could inhibit platelet aggregation and they called it "PG-X," [131]. PG-X later became known as PGI_2. It is 30 times more potent than any other known antiaggregatory agent [131]. PGI_2 is mainly produced by vascular endothelia cells. PGI_2 is a potent platelet inhibitor and a strong vasodilator. It counteracts with TXA_2 and plays a key role in maintenance of cardiovascular homeostasis related to vascular damage. Similar to TXA_2, PGI_2 also achieves its function through a paracrine and autocrine signaling cascade by binding to G-protein coupled prostacyclin receptor (IP receptor) on target cells including platelets and endothelial cells. To inhibit platelet aggregation, PGI_2 binds to IP receptor on platelets. The receptor activating process turns on adenyly cyclase to produce cAMP, which inhibits platelet activation and aggregation to counteract increased cytosolic calcium levels induced by TXA_2. To inhibit vasoconstriction, PGI_2 binds to IP receptors on vascular smooth muscle cells. Receptor activation turns on adenyly cyclase to increase cAMP levels in the cytosol. cAMP then activates protein kinase A (PKA), which continues the cascade by phosphorylating and inhibiting myosin light-chain kinase, which leads to smooth muscle relaxation and consequently vasodilatation.

In addition to arachidonic acid, 5,8,11,14,17-eicosatetraenoic acid (EPA) is also a substrate for thromboxane synthase and prostacyclin synthase. EPA produces TXA_3 and PGI_3. TXA_3 has less effect on platelet aggregation than TXA_2, but PGI_3 has equal activity as PGI_2 in inhibiting platelet aggregation and promoting vasodilatation.

Placental trophoblasts are a particularly rich source in the cyclooxygenase substrate arachidonic acid. Arachidonic acid is involved in cell membrane biosynthesis and represents a large portion of the total syncytiotrophoblast lipid [132]. Syncytiotrophoblasts are capable of incorporating arachidonic acid into both the brush border membrane (BBM) and basal plasma membrane compartments [133]. Trophoblast cells produce both TXA_2 and PGI_2. Placenta production of prostacyclin-like substances was first described by Myatt and Elder in 1977 [134]. They found a substance produced by human placenta that was capable of inhibiting platelet aggregation and discovered that this substance was heat sensitive [134]. Later, it was found that PGI_2 production (measured by its stable metabolite 6-keto-PGF1α), was significantly reduced in placentas delivered by women complicated with preeclampsia [135]. In contrast, production of TXA_2 was significantly increased [136,137]. The imbalance of increased TXA_2 production and decreased PGI_2 production in the preeclamptic placenta is believed to play a significant role in inducing placenta vasoconstriction in preeclampsia. The reason for altered TXA_2 production in placentas is not clear, but oxidative stress is considered one of the possible mechanisms that regulate TXA_2 and PGI_2 production. Study has shown that hypoxia not only promotes TXA_2 and PGI_2 productions, but also increases phospholipase A_2 production [138], which indicates that increased phospholipase A_2 release/activity can liberate further arachidonic acid from trophoblast membrane phospholipids to promote TX production. This results in an imbalance of increased TXA_2 and decreased PGI_2 production, consequently driving vasoconstriction of placental vessels in preeclampsia [138].

Consistent with the altered TXA_2 and PGI_2 productions in preeclamptic placentas, the ratio of TXA_2 levels and PGI_2 levels in maternal circulation are also increased in women with preeclampsia, and this increase is linked to the severity of disease [139]. The evidence of imbalanced thromboxane and prostacyclin in both the maternal circulation and the placenta compartment provides the rationale for the "low-dose aspirin therapy" for prevention of preeclampsia [140,141]. Aspirin is a nonsteroidal antiinflammatory drug and a nonselective inhibitor of cyclooxygenase. Aspirin inhibits both the cyclooxygenase-1 (COX-1) and cyclooxygenase-2 (COX-2) isoenzymes. By inhibiting the COX enzyme to limit the production of the precursor of thromboxane within platelets, aspirin irreversibly blocks the formation of TXA_2 in target cells and consequently inhibits platelet aggregation. This anticoagulant property makes aspirin useful for reducing the incidence of cardiovascular diseases including heart attacks. Although results from low-dose aspirin therapy clinical trials in preventing preeclampsia are inconsistent, timing and dose of the aspirin used in the various clinical trials and the heterogeneity of preeclampsia, early versus late onset, and/or with or without other clinical complications, may also contribute to the unsatisfactory outcomes [142]. Nonetheless, altered arachidonic acid-cycloxygenase-thromboxane pathway regulation in the placenta is widely believed to play an important role in the pathophysiology of preeclampsia.

We examined PGIS, TXS, and their receptors IP and TP expressions in placental tissues at different gestational ages in normal pregnancy and found that PGIS is strongly expressed in both cyto- and syncytiotrophoblasts in the first-trimester placenta. PGIS expression is reduced in advanced gestation (Figure 8.5). TXS is expressed in syncytiotrophoblasts throughout pregnancy. Both PGIS and TXS are expressed in the villous core fetal vessel endothelium in third-trimester placentas. Compared to TP receptor, IP receptor is strongly expressed in trophoblasts and fetal vessels in the first- and second-trimester placentas. Fetal vessel endothelium expresses both IP and TP receptors in the third-trimester placentas (Figure 8.5). The pattern for IP receptor over TP receptor expressions in trophoblasts and villous core vessels suggests that PGI_2 mediated vasodilatory activity is dominant in the placenta during normal pregnancy.

Other than the anticoagulation and vasodilation properties of PGI_2, recent studies also show that PGI2 plays a role in angiogenesis, and is involved in proinflammatory and/or antiinflammatory responses in endothelial cells. For instance, an *in vitro* study has shown that prostacyclin analogs could stimulate VEGF production by human lung mesenchymal cells/fibroblasts [143]. Kamio et al. studied prostacyclin effects on human lung fibroblasts using the prostacyclin analogs iloprost and beraprost [143], and found increased VEGF mRNA expression and protein release in cells treated with iloprost and beraprost. This prostacyclin-analog induced VEGF expression and production could be blocked by the adenylate cyclase inhibitor SQ-22536 and by a protein kinase A (PKA) inhibitor KT-5720 [143], indicating that prostacyclin downstream effects on VEGF is mediated

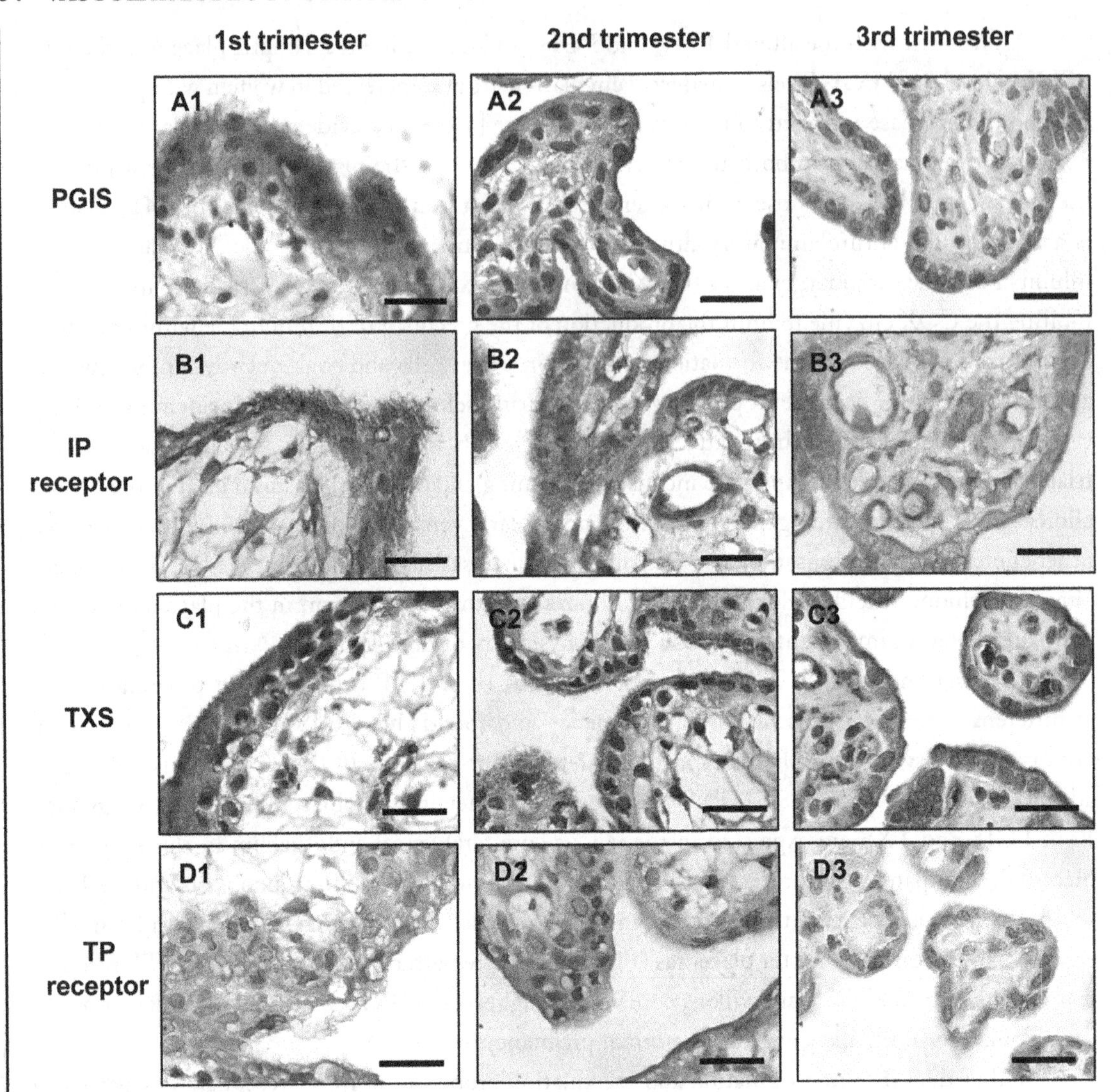

FIGURE 8.5: Immunostaining of prostacyclin synthase (PGIS), IP receptor, thromboxane synthase (TXS), and TP receptor in first-, second-, and third-trimester placentas. A: PGIS; B: IP receptor; C: TXS; and D: TP receptor. 1: First-trimester placentas; 2: second-trimester placentas; and 3: third-trimester placentas. Please note PGIS is strongly expressed in both cyto- and syncytiotrophoblasts in the first-trimester placenta. PGIS expression is reduced in advanced pregnancy. TXS is expressed in syncytiotrophoblasts throughout pregnancy. Both PGIS and TXS are expressed in the villous core fetal vessel endothelium in the third-trimester placentas. Compared to TP receptor, IP receptor is strongly expressed in the first- and second-trimester placentas. Fetal vessel endothelium expresses both IP and TP receptors in the third-trimester placentas. Bar = 25 micron.

via cAMP-activated PKA signaling cascade [143]. The finding of colocalizatioon of PGI_2 synthase with caveolin-1 in endothelial cells also suggests a potential angiogenic function of PGI_2 [144]. It is known that caveolin-1 expression is critical for VEGF-induced angiogenesis [145]. Both cytotrophoblasts and syncytiotrophoblast express caveolin-1 [146] and PGI_2 synthase and IP receptor are strongly expressed in the first-trimester placental cytotrophoblasts (Figure 8.5). It is highly likely that cytotrophoblasts-derived angiogenic factors are fundamental to stimulating villous core stromal vasculogenesis. The positive staining of PGI_2 synthase in the villous stromal cells [147] also suggests that prostacyclin derived from stromal mesenchymal cells may play a role in a bidirectional signaling network between the mesenchymal and vascular cells to promote vasculogenesis and angiogenesis during placenta development.

8.3 ENDOTHELIN-1 AND ITS RECEPTORS

Yanagisawa et al. found that a peptide originally derived from the supernatant from porcine aortic endothelial cells has a potent vasoconstrictive effect on porcine coronary artery strips [148]. The peptide was then purified and cloned from endothelial cells and named endothelin because it was derived from endothelial cells [148]. Endothelin has three isoforms, endothelin-1, -2, and -3. Endothelin-1 (ET-1) is the most potent and long lasting vasoconstrictor known, being 100 times more potent than noradrenaline [148]. Mature ET-1 is a 21-amino acid peptide, and it is a main member of the endothelin peptide family [149]. Endothelins has two receptors, ETA and ETB. ET-1 and -2 bind to ETA and ETB, while ET-3 only binds to ETB. ETs have been demonstrated to play important roles in cardiovascular diseases including hypertension, atherosclerosis, diabetes, and renal diseases [150].

In the full-term human placenta, the immunoreactivity of ET-1 is localized to endothelial cells of capillaries of placental microvilli, small- and medium-sized arteries and veins as well as placental syncytiotrophoblasts [151,152]. Studies also confirmed a broad distribution of ET-1 expression in placentas throughout gestation in which ET-1 expression is increased along with gestational age through the first trimester to full term [153]. Trophoblasts produce/release ET-1 [152,154]. Human placenta also expresses ETA and ETB receptors. We found that ETB is strongly expressed in placental trophoblasts and its expression is increased along with gestational age (Figure 8.6). In contrast, ETA is weakly expressed in placental trophoblasts. Although the reason for the differential expression of ETA and ETB in placental trophoblasts is not clear, an *in vitro* binding assay has clearly shown that ET is capable of binding to the isolated trophoblast membranes [155]. ET-1 is also involved in trophoblast invasion and differentiation of trophoblast cells isolated from first-trimester placentas [156].

The potential role of placenta-derived ET-1 in placental vessel vasoconstriction was studied in a vessel ring organ bath perfusion model [124]. It was found that conditioned medium derived from tissue culture of preeclamptic placentas could induce constriction of the chorionic plate artery ring of

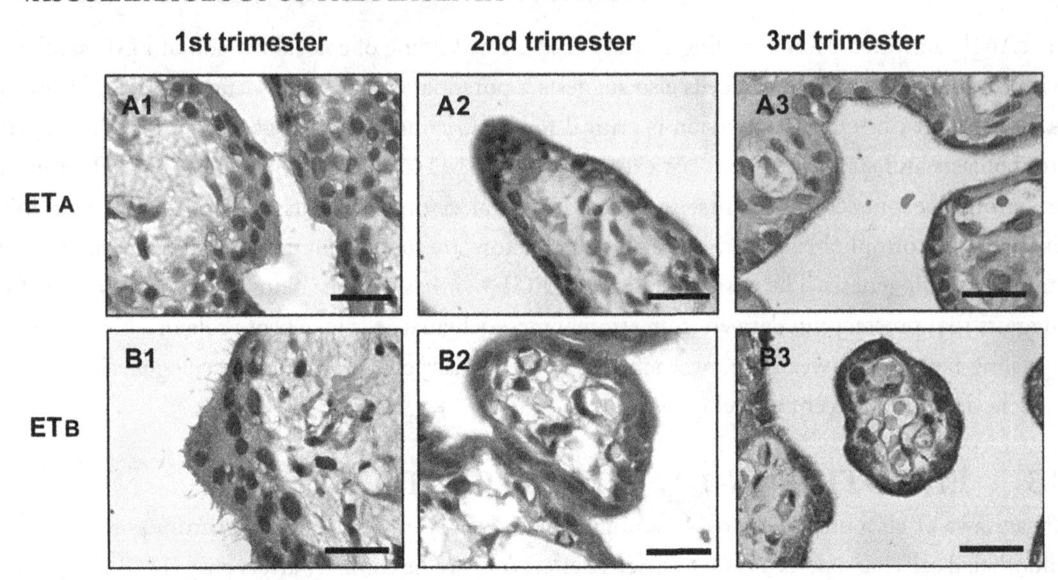

FIGURE 8.6: Immunostaining of ET_A and ET_B in first-, second-, and third-trimester placentas. A: ET_A; and B: ET_B. 1: First-trimester placentas; 2: second-trimester placentas; and 3: third-trimester placentas. ET_B is strongly expressed in placental trophoblasts and its expression is increased along with gestational age. ET_A is only weakly expressed in placental trophoblasts. Bar = 25 micron.

placentas from noncomplicated pregnancies and this vasoconstrictive effect could be attenuated by both ET_A and ET_B receptor antagonists [124], suggesting that ET_A and ET_B receptors are present in chorionic plate vessels and ET-1 produced by trophoblasts could induce vasoconstriction in the placenta [124]. Consistent with our organ bath perfusion study, Wilkes et al., also found that ET-1 could induce a sustained dose-dependent increase in perfusion pressure in umbilical cord artery and vein [157]. Paradoxically, ET-1 gene expression found no difference between placentas from normal and preeclamptic pregnancies [158]. Although the reason for this is not clear, the density difference between ET_A and ET_B receptors on placental trophoblasts could be an explanation. Activation of ET_A receptors on vascular smooth muscle cells induces vasoconstriction and activation of ET_B receptors on endothelial cells induces vasodilatation. Down-regulation of ET_A receptor expression but relatively increased ET_B receptor expression were found in preeclamptic placentas [158,159]. The increased ET_B receptor expression could be a compensatory effect that accounts for the partial protective effect during preeclampsia [158,159].The protective effect of ET-1 mediated by ET_B receptors is further supported by a study showing that ET-1 attenuated apoptosis of trophoblasts from full-term human placentas [160]. However, functions other than vasoactivity of ETs and their receptors are largely unknown in trophoblasts.

8.4 NITRIC OXIDE

Nitric oxide (NO) synthesis is catalyzed by nitric oxide synthase (NOS), which converts L-arginine to L-citruline with NO as a free radical by-product. There are three NOS isoforms, iNOS, eNOS, and nNOS, and the predominant constitutive NOS isoform within the placenta is eNOS. The biological effect of NO is mediated by two pathways: (1) the cGMP-dependent pathway involves activation of the NO-sensitive soluble form of guanylyl cyclase (sGC). This enzyme is responsible for cyclic-guanosine-monophosphate (cGMP) generation and protein kinase-G (PKG) activation; and (2) the cGMP-independent pathway involves the signaling molecules such as Ca^{2+}-activated and ATP-activated K^+ channels. Protein *S*-nitrosylation has been suggested as the possible underlying mechanism [161,162].

NO plays an important role, along with other vasodilators PGI_2, natriuretic peptides, and endothelial-derived hyperpolarizing factor (EDHF), in uteroplacental vascular adaptation during pregnancy. Strong eNOS activity is found in the vascular endothelium of the umbilical cord, chorionic plate, and stem villous vessels. *De novo* NO produced by the endothelium is secreted adluminally to adjacent vascular smooth muscle cells (Figure 8.7). The adluminally directional release of NO supports the role of NO in the transformation of the uterine arteries and favors its participation in maintaining the uteroplacental vasorelaxation.

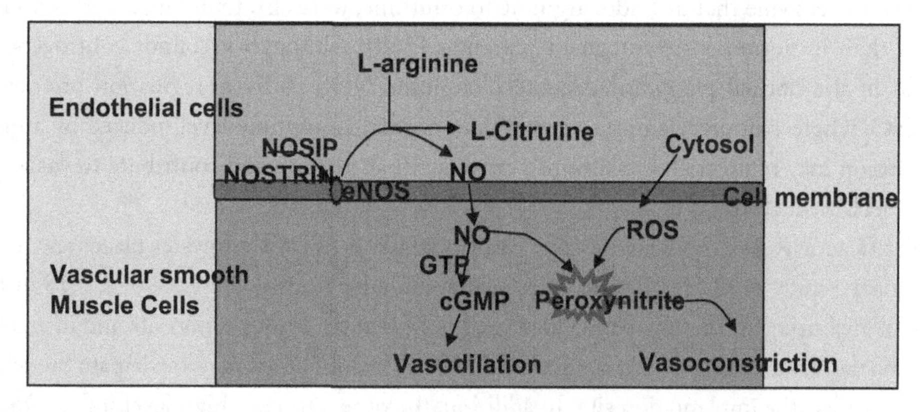

FIGURE 8.7: NO pathway. The constitutive nitric oxide synthase (eNOS) is expressed in vessel endothelium. When acting on its substrate L-arginine, nitric oxide (NO) is produced as a by-product to generate L-Citruline. NO diffuses to the underlying vascular smooth muscle cells (VSMCs) where they act on GTP to produce cGMP, which is capable of dilating blood vessels. When excess ROS is present, it reacts with NO to form free radical peroxynitrite, which is able to induce oxidative stress and subsequently vasoconstriction in placental vessels. Two eNOS regulating proteins have been identified: NOSTRIN and NOSIP. They may be involved in protein network controlling activity, trafficking, and targeting of eNOS.

The importance of NO is highlighted by the improvement of the local and systemic perfusion changes in hypertensive pregnant patients supplemented with L-arginine. In patients with mild preeclampsia, administration of NO donor glyceryl trinitrate significantly reduces the blood flow resistance in fetoplacental circulation as examined by Doppler monitoring [163]. Consistently, inhibition of NOS activity on isolated human stem villous arterioles increases umbilical vascular resistance with reduced blood flow [164]. Additional evidence comes from the experiments showing that inhibition of sGC increases the perfusion pressure of the human fetoplacental circulation [165]. Substantial evidence from animal studies has also shown a key role of NO in shear stress mediated vasodilation in the uteroplacental vasculature [166]. Thus, it is clear that NO is an important endogenous dilator of the fetal vessels in the placenta.

Several studies have found that eNOS activity and NO end-products (nitrites and nitrates) are significantly high in villous homogenate and eNOS is intensively expressed in endothelium of stem villous vessels and small arteries of terminal villi in preeclamptic placentas compared to those in normal placentas [167,168]. These observations suggest that increased placental NO production may represent a compensatory mechanism to offset the pathologic effects in preeclampsia. In other studies, L-arginine levels, a substrate for endogenous NO synthesis in umbilical blood and in villous tissues, were found lower, but nitrotyrosine staining, a marker of peroxynitrite, was stronger in preeclamptic than in normotensive pregnancy [169]. Gene expression and protein tissue content of arginase II, the enzyme that degrades arginine to ornithine, were also found higher in preeclamptic placentas than in normotensive pregnant placentas [169]. Although it sounds controversial, it is clear that in the normal placenta, adequate L-arginine/NOS pathway regulation orients eNOS toward NO, whereas in preeclampsia a lower than normal L-arginine level induced by arginase II overexpression may redirect eNOS toward peroxynitrite formation and contribute to vasoconstriction in preeclampsia [169].

eNOS activity has also been described in trophoblasts of early trimester placentas. Immunohistochemistry study of the first-trimester placenta demonstrates the presence of eNOS in the cell columns of anchoring villi and in extravillous trophoblasts at the implantation site and in villous syncytiotrophoblasts, indicating that *in situ* produced NO by trophoblasts may participate the relaxation of vascular wall at the implantation site. In addition, the observation of high levels of eNOS immunoreactivity in the intermediate trophoblasts in complete hydatidiform mole and in choriocarcinoma compared to placentas with nontrophoblast disease suggests that eNOS is also capable of stimulating trophoblast proliferation [170,171].

Recently, two eNOS regulating proteins, an eNOS traffic inducer (NOSTRIN) and an eNOS-interacting protein (NOSIP) were reported [172]. NOSTRIN is found to colocalize extensively with eNOS at the plasma membrane of confluent human umbilical venous endothelial cells, and eNOS-NOSTRIN interactions are confirmed in both *in vitro* and *in vivo* experimental studies [172]. Furthermore, NOSTRIN overexpression could induce a profound redistribution of eNOS from the plasma membrane to vesicle-like structures matching the NOSTRIN pattern and at the same time led to a significant inhibition of NO release [172]. These data indicate that NOSTRIN contributes to the intricate protein network controlling activity, trafficking, and targeting of eNOS. Although, at present, no information is available regarding NOSIP and/or NOSTRIN in placental trophoblasts or villous core vessels, it is expected that NOSIP and NOSTRIN may regulate eNOS translocation and trafficking in placenta vasculature.

8.5 CHYMASE

Chymase is a chymotrypsin-like serine protease. Chymase has no enzymatic activity in the normal state but is activated immediately upon release into the extracellular matrix. Chymase activation is associated with many pathophysiological conditions and plays an important role in vessel reactivity because it is involved in both angiotensin II and endothelin biosynthesis. Like ACE, chymase converts angiotensin I to angiotensin II by hydrolyzing bonds between Phe_8-His_9. Chymase is considered a potent non-ACE angiotensin II generating enzyme and found to be responsible for approximately 70–80% angiotensin II generated in the human heart tissue [173]. Chymase is also an endothelin-forming enzyme. It cleaves the Trp_{21}-Val_{22} bond of big-ET-$1_{(1-39)}$ forming ET-$1_{(1-21)}$ and the Tyr_{31}-Gly_{32} bond, resulting in the formation of ET-$1_{(1-31)}$ [174]. Both ET-$1_{(1-21)}$ and ET-$1_{(1-31)}$ bind to ET-A receptor on vascular smooth muscle cells. Angiotensin II and ET-1 are potent vasoconstrictors, thus chymase plays a fundamental role in hypertension and atherosclerosis. The major pathophysiological effects of chymase are shown in Figure 8.8.

The chymase gene is found in the placental trophoblasts and the open reading frame of the chymase gene in trophoblasts is 100% homologous to that reported in the human heart tissue and mast cells [175–177]. Importantly, chymase expression is up-regulated and chymase activity is increased in trophoblasts of placentas from women with preeclampsia [175], suggesting that chymase may

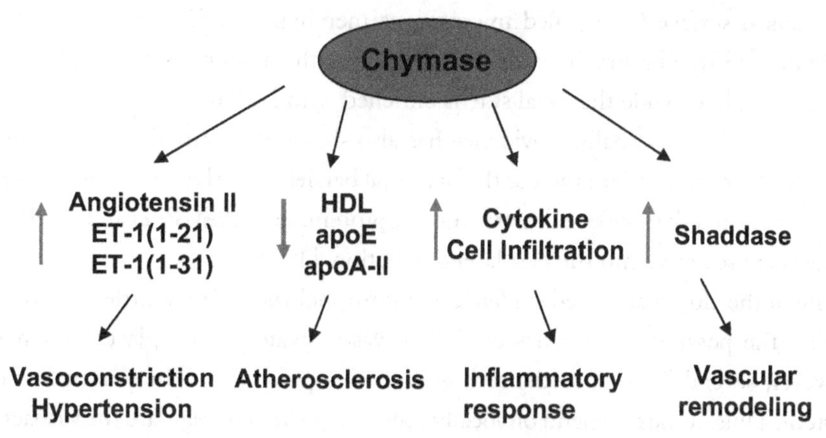

FIGURE 8.8: Potential role of chymase in the cardiovascular system.

contribute to the altered RAS system, as well as to ET-1 production in preeclamptic placentas. The placental chymase is mainly localized in syncytiotrophoblasts [175]. The evidence of chymostatin-blocking effects on preeclamptic placenta-derived factor induced vasoconstriction of chorionic plate arteries and placental vessel smooth muscle cells further indicates the role of trophoblast-derived chymase mediated placenta vasoactivity in preeclampsia [124,178].

In addition to vasoconstrictor generating activity, chymase also plays a role in atherosclerosis by degrading HDL and inhibiting the apolipoprotein-mediated removal of cholesterol [179,180]. Chymase is considered to be an inflammatory protease [181]. It triggers the production of cytokines and chemokines, which stimulate the infiltration of inflammatory cells. In mast cells, degranulation not only release chymase but also biogenic amines, TNFα, and serglycin proteoglycans, as well as various lysosomal enzymes [182]. Similarly, increased chymase activity is associated with endothelial activation and increased endothelial adhesion molecule expression [183]. Although the inflammatory activity of chymase is not clear in the placenta, it is expected that the role of chymase in the placenta is beyond the ACE and ECE properties, since the sheddase activity of chymase was found to be associated with increased soluble VEGF receptor-1 (sFlt-1 release) in preeclamptic placentas [184].

8.6 ROLE OF TROPHOBLASTS IN REGULATION OF PLACENTAL VASOACTIVITY

The syncytio layer of trophoblasts has a unique hemonochorial structure: the apical surface of trophoblasts bathes in the maternal blood in the intervillous space, while the basal membrane (BM) is close to the villous stroma and the fetal vasculature. This anatomic feature of placental trophoblasts makes them an ideal candidate to mediate the signal from the maternal source and to regulate the function of underlying tissues, including villous core fetal vessels. Placental trophoblasts are polarized epithelial cells. These cells have diverse properties in chemical and biochemical compositions between the apical surface (also called microvillous membrane/MVM or brush border) and the basal membrane [185]. The brush border is enriched with membrane-bound alkaline phosphatase and 5' nucleotidase, while the basal side is enriched with Na^+-K^+ ATPase and F(-)-stimulated adenylate cyclase [186]. Compelling evidence has also shown differential characteristics of MVM from BM in transporting nutrients across the placental barrier from the mother to fetus [187]. It has also been demonstrated that >90% of matrix metalloproteinase-2 (MMP-2) and MMP-9 produced by trophoblasts are secreted into the basolateral direction. [188].

In spite of the aforementioned evidence about trophoblast polarity, little information is available regarding the possible polarized secretion of vasoactivators from placental syncytiotrophoblasts. However, several lines of evidence highlight this possibility. The placenta is devoid of a nervous system. Thus, it must depend on locally produced factors to regulate the vasoactivity within the placenta. This notion is best illustrated in the case of preeclampsia. Preeclampsia is characterized by systemic vasoconstriction. In particular, the uteroplacental vasculature is prominently affected. The underlying mechanism has been attributed to the imbalance of increased vasoconstrictors and decreased vasodilators exemplified by TXA_2 and PGI_2. What seems missing is an explanation for how the short-lived TXA_2 secreted from the placental trophoblasts could mediate constriction of placental vessels. Chorionic arteries from preeclamptic placentas displayed a greater contractility than that from normal placentas either to the vasoconstricting agent KCL or placenta conditioned medium [124], which were corroborated by corresponding inhibitors to thromboxane, endothelin, and angiotensin II/chymase. Based on the anatomical reciprocation between placental syncytiotrophoblasts and placental vessels, it is logical to reason that trophoblasts play a significant role in affecting vasoactivity in the placenta.

We recently found that predominant basal directional release of vasoconstrictor(s) is very likely a feature of syncytiotrophoblasts in regulation of placental villous core vessel activity during pregnancy, especially in preeclampsia [147]. TXS expression is increased in placental trophoblasts in preeclampsia [147]. Trophoblasts release TXA_2 to both the apical and basal directions, and the bidirectional release of TXA_2 is increased in trophoblasts from preeclamptic placentas [147]. Most interestingly, apical exposure of trophoblasts to arachidonic acid, a substrate of cyclooxygenase, results in an increase in TXA_2, but not PGI_2, releases in both the apical and basal directions. However, apical exposure of cyclooxygenase inhibitor aspirin could only inhibit the bidirectional release of TXA_2 in trophoblasts from normal placentas but not in trophoblasts from preeclamptic placentas. These observations suggest that arachidonic acid induced increase in bidirectional release of TXA_2 is mediated via cyclooxygenase in the normal trophoblasts, whereas TXS probably plays a dominant role in producing TXA_2 and regulating the bidirectional release of TXA_2 in the preeclamptic placenta, since upregulation of TXS is seen in preeclamptic placentas [147].

Although at the present time, only thromboxane has been shown to have a polarized secretion pattern from the placental syncytiotrophoblasts, it can be anticipated that other vasoactivators might also present a similar pattern. Chymase expression is upregulated in placental trophoblasts in preeclampsia [175]. An *in vitro* study has shown that chymotrypsin/chymase promotes basal directional release of angiotensin II by endothelial cells [189]. In the endothelium, ET-1 is also predominantly released abluminally toward the vascular smooth muscle, suggesting a paracrine role in the regulation of vascular smooth muscle contraction [190,191]. Chymase acts as both ACE and ECE to promote angiotensin II and ET-1 productions. Therefore, increased trophoblast chymase activity is likely to contribute to the increased placental vasoconstriction mediated by basal directional release of both angiotensin II and ET-1 in preeclampsia.

• • • •

CHAPTER 9

Lymphatic Phenotypic Characteristics of the Human Placenta

The lymphatic system is complementary to the circulating vascular blood system. Lymphatic vessels are found in almost all organs and tissues except the brain. They act as a reservoir for plasma and other substances, including cells and interstitial fluid, that leaked from the vasculature and they transport lymph and interstitial fluid back from the tissues to the circulatory system. Without a functioning lymph system, interstitial fluid and lymph cannot be effectively drained. Lymph vessel blockage and abnormal lymphatic vessel function would result in tissue swelling (edema).

During pregnancy, maternal uterine blood vessels undergo dramatic vascular remodeling. Uteroplacental blood flow increases progressively during pregnancy with estimates ranging from 600–700ml/min near term. The increased uteroplacental blood flow accommodates the growth and development of the placenta and the fetus. Recent work indicates in addition to uterine blood vessel adaptation, dramatic changes in the uterine lymphatic vessel system are also involved in the vascular remodeling during pregnancy [192]. The uterine lymphatic system functions in diverse processes involving interstitial fluid homeostasis and adaptive immunity [193].

Although the lymphatic system plays an important role in vessel remodeling and fluid homeostasis in the uterus during pregnancy, no lymphatic vessel and/or capillary is found in the placenta. However, recent developments have shown that numerous lymphatic markers are expressed in human placental tissue, including VEGF-C, VEGF-D, prospero-related homeobox-1 (Prox-1), lymphatic vascular endothelial hyaluronan receptor-1 (LYVE-1), VEGF receptor -3 (VEGFR-3/Flt-4), and D2-40. Very interestingly, the expression and localization of these lymphatic markers are compartmentally different within the villous tissue (Table 9.1).

As discussed earlier, placental trophoblasts are unique syncytialized epithelial cells. They form a syncytial layer and cover the entire surface of villous core fetal tissues. They are bathed in the maternal blood within the intervillous space and are the barrier of the maternal-fetal interface. The nature of placental trophoblasts—their direct contact to maternal circulating components—makes these cells uniquely behave as "endothelial cells," because they express many endothelial markers. Although no lymphatic vessel is found in the placenta, studies by our group did discover

TABLE 9.1: Lymphatic markers expressed in the human placenta

MARKER	LOCALIZATION	DETECTION METHODS
VEGF-C	Villous core endothelial cells	mRNA expression by PCR
VEGF-D	Trophoblasts Villous core endothelial cells	mRNA expression by PCR
VEGFR-3/Flt-4	Trophoblasts Villous core endothelial cells	Immunostaining, Western blot mRNA expression by PCR
Prox-1	Trophoblasts Villous core endothelial cells	mRNA expression by PCR
LYVE-1	Trophoblasts Villous core endothelial cells	mRNA expression by PCR Immunostaining
D2-40 (Podoplanin)	Stroma	Immunostaining, Western blot

that many lymphatic markers are expressed in the placenta [67]. For example, we found in the human placenta expression of lymphatic vascular hyaluronan receptor LYVE-1 in, which appears to compensate for the absence of CD44 in the placenta [67]. Other than VEGF, Flt-1, and VE-cadherin [47], and so forth, similar to villous core fetal endothelial cells, syncytiotrophoblasts express VEGFR-3/Flt-4. In the systemic lymphative vessels, VEGF-C and VEGF-D and their receptor VEGFR-3 function for lymphangiogenesis. Expression of these lymphatic marker molecules point out lymphatic phenotypic characteristics of placental trophoblasts, although particular role of these molecule function has not been elucidated yet.

Among these lymphatic markers expressed in the placental villous tissue, D2-40 (podoplanin) is found so far to be the only one expressed in the villous stroma. D2-40 reacts with an O-linked sialoglycoprotein directly against human podoplanin, a mucin-type transmembrane protein originally reported in lymphatic endothelial cells [194]. It also reacts with oncofetal membrane antigen M2A in fetal testes and seminomas [195]. It is believed that D2-40 immunostaining may serve as a marker for increased risk of lymphatic invasion. We examined both D2-40 and M2A expressions in the human placentas, and found that D2-40 was strongly expressed in the villous stoma throughout gestation in the human placenta, whereas M2A was not expressed. The differential expression of D2-40 and M2A antigen suggests the phenotypic difference of these two lymphatic markers in the placental tissue. We further found compartmental differences for D2-40, CD31, and VEGFR-3 distributions in the placental villous tissue. D2-40 is localized in the stroma viewing as a network plexus pattern; CD31 is expressed only in villous core fetal vessel endothelium, and VEGFR3 is seen in both trophoblasts and fetal vessel endothelium. The diffused network pattern of D2-40 expression in villous tissue is confirmed throughout gestation in first-, second-, and third-trimester placental tissue sections (Figure 9.1).

Placental villous tissue contains abundant mesenchymal and matrix channels, especially in the immature intermediate villi. Placental villous stroma is a unique channel-like structure [196]. It consists of a network of cells and fibers with fetal vessels. Within the stroma, it forms a fluid system compartment with Hofbauer cells suspended in the interspaces [196]. Mesenchymal and matrix channels provide a path for the Hofbauer cells to patrol the villous core. The localization of D2-40 in the stroma suggests that a lymphatic-like conductive network might exist in the human placenta. In addition to lymphatic endothelial cells, podoplanin is also expressed in germ cells, mesothelial cells, stromal reticular cells, and many types of tumor tissues [197]. Interestingly, we recently found that D2-40 expression is significantly reduced in placentas from women with preeclampsia [198]. Sialoglycoproteins are widely distributed in cell membrane and serum. The protein-bounded carbohydrates contribute to the physicochemical properties of glycoproteins, protect glycoproteins from

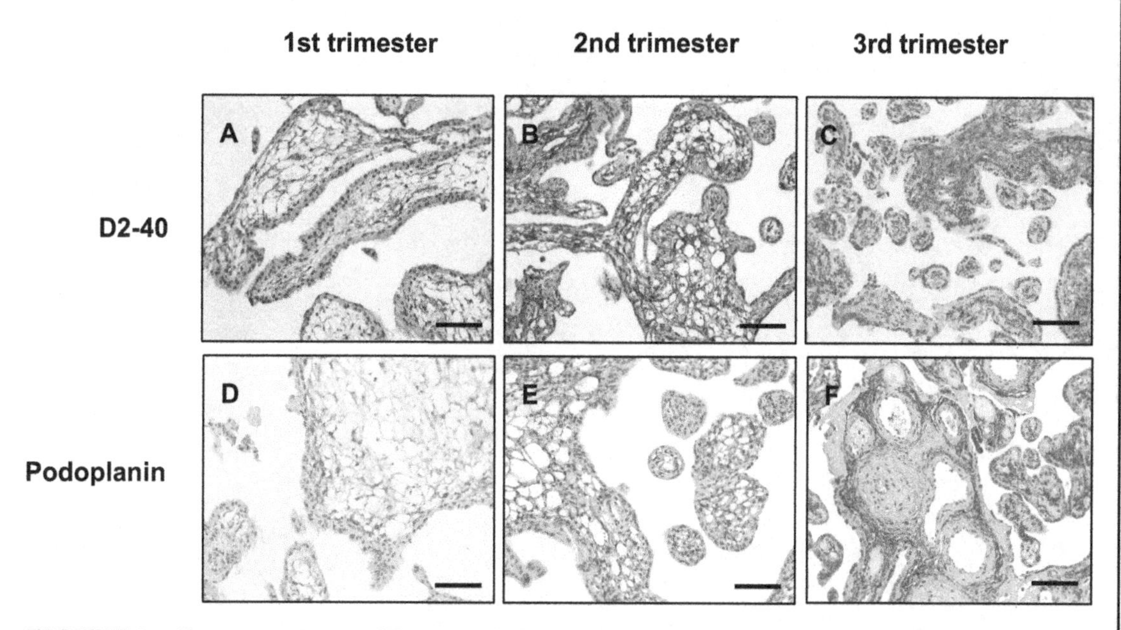

FIGURE 9.1: Immunostaining of D2-40 and podoplanin in first-, second-, and third- trimester placental tissue sections. Note that the network pattern of D2-40 staining is seen in villous stroma throughout pregnancy. Podoplanin expression exhibits the same pattern as D2-40. Bar = 100 μm.

proteolytic attacks, and function in signal recognition in cell-cell and ligand-receptor interactions. Although little is known regarding the biological functions of sialylated glycoprotein in the placenta, strong D2-40 expression in the villous stoma suggests that podoplanin may play an important role in the interstitial fluid balance within the villous stroma. Reduced D2-40 expression may lead to failure of interstitial fluid conduction, increased interstitial fluid accumulation and pressure, and may also affect Hofbauer cells trafficking in the villous core stroma. Poor placental tissue perfusion in preeclampsia might thus provoke higher perfusion pressures, which are pathognomonic for preeclampsia. However, whether reduced D2-40 sialoglycoprotein expression and/or dysregulation of the protein function contribute to improper trophoblast invasion and abnormal placental angiogenesis needs further investigation.

• • • •

CHAPTER 10

Placental Tissue and Cord Blood Stem Cells

The human placenta and cord blood are rich in hematopoietic progenitor and hematopoietic stem cells (HSCs), which give rise to all the blood cell types including myeloid (monocytes and macrophages, neutrophils, basophils, eosinophils, erythrocytes, megakaryocytes/platelets, and dendritic cells) and lymphoid lineage (T-cells, B-cells, and NK cells) cells. Since the first successful umbilical

TABLE 10.1. Differentiation potential of human placenta and amniotic fluid mesenchymal stem cells

CELL TYPE	EXPRESSION MARKER AND/OR PHENOTYPE
Osteogenic	von Kossa stain, alkaline phosphatase, osteocalcin
Chondrodenic	collagen type II immunochemical detection
Adipogenic	Oil Red-O stain
Cardiomyogenic	MyoD, desmin
Skeletal myogenic	myosin
Neurogenic	Release of catecholamine, acetylcholine, neurotropic factors, activin, noggin, nurofilament
Pancretic	PDX-1, PAx-6, NKx2.2, insulin, glucagon
Hepatic	Albumin, HNF-4, A1AT
Angiogenic	Flt-1, KDR, ICAM-1, CD34
Multi-lineage SP cells*	HLA I(-)/II(-), HLA I(+)/II(-)

* Side population cells

cord blood transplants in children with Fanconi anemia [199], cord blood banking and the therapeutic use of cord blood stem cells have grown quickly in the last two decades. Like stem cells from bone marrow, umbilical cord blood hematopoietic stem cells have been used to treat various genetic disorders including leukemia, certain cancers, and some inherited disorders [200–202].

In addition to hematopoietic progenitors and HSCs, the placenta also enriches mesenchymal stem cells (MSCs). MSCs have a broad differentiation potential [203]. MSCs from fetal membranes and placental tissue are able to differentiate *in vitro* into multiple lineage cell types including osteoblasts, chondrocytes, myocytes, adipocytes, and endothelial cells [203–207]. Recent animal studies also showed that MSCs could be differentiated into beta-pancreatic islet cells and MSCs administration result in beta-pancreatic islet regeneration and prevent renal damage in diabetic animals [208,209]. Table 10.1 shows a list of differentiation potential of human placenta and amniotic fluid mesenchymal stem cells. Mesenchymal stem cells are well tolerated by the host and can therefore avoid allorecognition. They interfere with dendritic cells and T-cells. Theses cells are able to secrete cytokines and generate a local immunosuppressive microenvironment [210].

Placental MSCs are easily accessible with more primitive molecular characteristics. They are amplifiable *in vitro*. The multiple-lineage differentiation potential of placental MSCs presents an invaluable future for clinical applications. To date, MSCs have been successfully isolated from human placental villous tissue, amniotic fluid, and fetal membrane. They offer a renewable source of cell replacement for antiaging therapy, regeneration medicine, and are used to treat various neurological and immune disorders. Placental stem cell therapy will be a promising answer for many of today's untreatable diseases in the years to come.

• • • •

CHAPTER 11

Summary

The placenta is the first organ to be formed during pregnancy. Although the placenta is a temporary organ connecting the mother and the fetus, it plays fundamental roles during pregnancy. The placenta controls oxygen and metabolite exchange, produces growth factors and hormones, and transfers nutrients to support fetal development. The placenta modifies maternal adaptation to pregnancy. Fetal survival and growth are also dependent on a well-established and functional placenta. However, healthy pregnancy outcomes to both the mother and fetus rely on normal functional placental trophoblasts and proper remodeling of uterine spiral arteries during the earlier stages of pregnancy. Physiological conversion of the uterine spiral arteries and adequate maternal blood supply to perfuse the placenta is key to a successful human pregnancy. In contrast, defective placentation, impaired trophoblast invasion, and failure of sufficient remodeling of spiral arteries are often the common features of adverse pregnancy outcomes including early pregnancy loss, intrauterine growth restriction, and preeclampsia. These adverse pregnancy outcomes in women associated with abnormal placental and vascular development during pregnancy are now recognized as a predisposing factor that makes a significant impact on the development of cardiovascular diseases later in life. Studies have shown that women with a history of abnormal placental syndromes are at additional risk of developing cardiovascular diseases and metabolic disorders in later life such as hypertension, ischemic heart diseases, stroke, and diabetes. In addition, abnormal placental function (increased vascular resistance, improper nutrient transport, and epigenetic gene imprinting) also has an impact on fetal adaptations central to programming. Although, in the past few decades, significant progress has been made in the field of placental vascular biology owing to advanced cellular and molecular technologies. Still, the placenta remains a fascinating and enigmatic organ. Many unknowns await answers. Therefore, placental research has always been an active area of investigation in the past, today, and will continue to be in the future.

• • • •

Acknowledgment

This work is supported in part by grants from the National Institute of Health, NHLBI (HL65997) and NICHD (HD36822). The authors wish to thank Dr. J. Steven Alexander for his careful reading and invaluable input for this work.

Acknowledgment

[illegible]

References

[1] Ellery PM, Cindrova-Davies T, Jauniaux E, Ferguson-Smith AC, Burton GJ. Evidence for transcriptional activity in the syncytiotrophoblast of the human placenta. *Placenta.* 2009;30(4):329–334.

[2] Jackson MR, Mayhew TM, Boyd PA. Quantitative description of the elaboration and maturation of villi from 10 weeks of gestation to term. *Placenta.* 1992;13(4):357–370.

[3] Lang I, Pabst MA, Hiden U, Blaschitz A, Dohr G, Hahn T, Desoye G. Heterogeneity of microvascular endothelial cells isolated from human term placenta and macrovascular umbilical vein endothelial cells. *Eur J Cell Biol.* 2003;82(4):163–173.

[4] Burrows TD, King A, Loke YW. Trophoblast migration during human placental implantation. *Hum Reprod Update.* 1996;2(4):307–321.

[5] Meekins JW, Luckas MJM, Pijnenborg R, McFadyen IR. Histological study of decidual spiral arteries and the presence of maternal erythrocytes in the intervillous space during the first trimester of normal human pregnancy. *Placenta.* 1997;18:459–464.

[6] Jaffe R, Jauniaux E, Hustin J. Maternal circulation in the first-trimester human placenta—myth or reality? *Am J Obstet Gynecol.* 1997;176(3):695–705.

[7] Cunningham FG, Leveno KJ, Bloom SL, Hauth JC, Gilstrap III LC, Wenstrom KD. Chapter 3. Implantation, embryogenesis, and placental development. Williams Obstetrics, 22nd Edition 2005:39–90.

[8] Gowland PA, Francis ST, Duncan KR, Freeman AJ, Issa B, Moore RJ, Bowtell RW, Baker PN, Johnson IR, Worthington BS. In vivo perfusion measurements in the human placenta using echo planar imaging at 0.5 T. *Magn Reson Med.* 1998;40(3):467–473.

[9] Benirschke K, Kaufmann P, Baergen R. Chapter 12. Anatomy and pathology of the umbilical cord. Pathology of the human placenta. 5th Edition. Ed: Benirschke K, Kaufmann P, Baergen R. 2006:380–451.

[10] Harold JG, Siegel RJ, FitzGerald GA, Satoh P, Fishbein MC. Differential prostacyclin production by human umbilical vasculature. *Arch Pathol Lab Med.* 1988;112(1):43–46.

[11] Karbowski B, Bauch HJ, Schneider HP. Functional differentiation of umbilical vein endothelial cells following pregnancy complicated by smoking or diabetes mellitus. *Placenta.* 1991;12:405.

[12] Giles W, O'Callaghan S, Read M, Gude N, King R, Brennecke S. Placental nitric oxide synthase activity and abnormal umbilical artery flow velocity waveforms. *Obstet Gynecol.* 1997;89(1):49–52.

[13] El Behery MM, Nouh AA, Alanwar AM, Diab AE. Effect of umbilical vein blood flow on perinatal outcome of fetuses with lean and/or hypo-coiled umbilical cord. *Arch Gynecol Obstet.* 2009 Nov 7. [Epub ahead of print].

[14] Castellucci M, kaufmann P. Chapter 7. Architecture of naormal villous trees. Pathology of the human placenta. 5th Edition. Ed: Benirschke K, Kaufmann P, Baergen R. 2006, 121–173.

[15] Castellucci M, Scheper M, Scheffen I, Celona A, Kaufmann P. The development of the human placental villous tree. *Anat Embryol.* (Berl). 1990;181(2):117–128.

[16] Castellucci M, Kosanke G, Verdenelli F, Huppertz B, Kaufmann P. Villous sprouting: fundamental mechanisms of human placental development. *Hum Reprod Update.* 2000; 6(5):485–494.

[17] Cross JC, Werb Z, Fisher SJ. Implantation and the placenta: key pieces of the development puzzle. *Science.* 1994;266:1508–1518.

[18] Crocker IP, Strachan BK, Lash GE, Cooper S, Warren AY, Baker PN. Vascular endothelial growth factor but not placental growth factor promotes trophoblast syncytialization in vitro. *J Soc Gynecol Investig.* 2001;8:341–346.

[19] Nelson DM, Crouch EC, Curran EM, Farmer DR. Trophoblast interaction with fibrin matrix. *Am J Pathol.* 1990;136:855–865.

[20] Torry DS, Ahn H, Barnes EL, Torry RJ. Placenta growth factor: potential role in pregnancy. *Am J Reprod Immunol.* 1999;41:79–85.

[21] Brosens IA, Robertson WB, Dixon HG. The role of the spiral arteries in the pathogenesis of preeclampsia. *Obstet Gynecol Ann.* 1972;1:177–191.

[22] Robertson WB, Brosens I, Dixon HG. The pathological response of the vessels of the placental bed to hypertensive pregnancy. *J Pathol Bacteriol.* 1967;93:581–592.

[23] DeWolf F, Robertson WB, Brosens I. The ultrastructure of acute atherosis in hypertensive pregnancy. *Am J Obstet Gynecol.* 1975;123:164–174.

[24] Burton GJ, Woods AW, Jaunlaux E, Kingdom JCP. Rheological and physiological consequences of conversion of the maternal spiral arteries for uteroplacental blood flow during human pregnancy. *Placenta.* 2009;30(6):473–482.

[25] Demir R, Kaufmann P, Castellucci M, Erbengi T, Kotowski A. Fetal vasculogenesis and angiogenesis in human placental villi. *Acta Anat* (Basel). 1989;136(3):190–203.

[26] Hofbauer J. Uber das konstante vorkommen bisher unbekannter zelliger Formelemente in der Chorionzotte der menschlichen Plazenta und uber Embryotrophe. *Wien. Klin. Wochenschr.* 1903;16:871–873.

[27] Khan S, Katabuchi H, Araki M, Nishimura R, Okamura H. Human villous macrophage-conditioned media enhance human trophoblast growth and differentiation in vitro. *Biol Reprod.* 2000;62(4):1075–1083.

[28] Cooper JC, Sharkey AM, McLaren J, Charnock-Jones DS, Smith SK. Localization of vascular endothelial growth factor and its receptor, flt, in human placenta and decidua by immunohistochemistry. *J Reprod Fertil.* 1995;105(2):205–213.

[29] Ahmed A, Li XF, Dunk C, Whittle MJ, Rushton DI, Rollason T. Colocalisation of vascular endothelial growth factor and its Flt-1 receptor in human placenta. *Growth Factors.* 1995;12(3):235–243.

[30] Demir R, Kayisli UA, Seval Y, Celik-Ozenci C, Korgun ET, Demir-Weusten AY, Huppertz B. Sequential expression of VEGF and its receptors in human placental villi during very early pregnancy: differences between placental vasculogenesis and angiogenesis. *Placenta.* 2004;25(6):560–572.

[31] Seval Y, Korgun ET, Demir R. Hofbauer cells in early human placenta: possible implications in vasculogenesis and angiogenesis. *Placenta.* 2007;28(8-9):841–845.

[32] Anteby EY, Natanson-Yaron S, Greenfield C, Goldman-Wohl D, Haimov-Kochman R, Holzer H, Yagel S. Human placental Hofbauer cells express sprouty proteins: a possible modulating mechanism of villous branching. *Placenta.* 2005;26(6):476–483.

[33] Cabrita MA, Christofori G. Sprouty proteins, masterminds of receptor tyrosine kinase signaling. *Angiogenesis.* 2008;11(1):53–62.

[34] Natanson-Yaron S, Anteby EY, Greenfield C, Goldman-Wohl D, Hamani Y, Hochner-Celnikier D, Yagel S. FGF 10 and Sprouty 2 modulate trophoblast invasion and branching morphogenesis. *Mol Hum Reprod.* 2007;13(7):511–519.

[35] Sims DE. Diversity within pericytes. *Clin Exp Pharmacol Physiol* 2000;27:842–846.

[36] Hirschi KK, D'Amore PA. Pericytes in the microvasculature. *Cardiovasc Res.* 1996;32:687–698.

[37] Challier JC, Kacemi A, Olive G. Mixed culture of pericytes and endothelial cells from fetal microvessels of the human placenta. *Cell Mol Biol.* 1995;41(2):233–241.

[38] Dye JF, Jablenska R, Donnelly JL, Lawrence L, Leach L, Clark P, Firth JA. Phenotype of the endothelium in the human term placenta. *Placenta.* 2001;22:32–43.

[39] Lang I, Schweizer A, Hiden U, Ghaffari-Tabrizi N, Hagendorfer G, Bilban M, Pabst MA, Korgun ET, Dohr G, Desoye G. Human fetal placental endothelial cells have a mature arterial and a juvenile venous phenotype with adipogenic and osteogenic differentiation potential. *Differentiation.* 2008;76(10):1031–1043.

[40] Rodesch F, Simon P, Donner C, Jauniaux E. Oxygen measurements in endometrial and trophoblastic tissues during early pregnancy. *Obstet Gynecol.* 1992;80(2):283–285.

[41] Jauniaux E, Watson A, Burton G. Evaluation of respiratory gases and acid-base gradients in human fetal fluids and uteroplacental tissue between 7 and 16 weeks' gestation. *Am J Obstet Gynecol.* 2001;184(5):998–1003.

[42] James JL, Stone PR, Chamley LW. The regulation of trophoblast differentiation by oxygen in the first trimester of pregnancy. *Hum Reprod Update.* 2006;12(2):137–144.

[43] Lyall F, Bulmer JN, Duffie E, Cousins F, Theriault A, Robson SC. Human trophoblast invasion and spiral artery transformation: the role of PECAM-1 in normal pregnancy, pre-eclampsia, and fetal growth restriction. *Am J Pathol.* 2001;158(5):1713–1721.

[44] Genbacev O, Joslin R, Damsky CH , Polliotti BM, Fisher SJ. Hypoxia alters early gestation human cytotrophoblast differentiation/invasion in vitro and models the placental defects that occur in preeclampsia. *J Clin Invest.* 1996;97:540–550.

[45] Genbacev O, Zhou Y, Ludlow JW, Fisher SJ. Regulation of human placental development by oxygen tension. *Science.* 1997;277:1669–1672.

[46] Fisher SJ, Roberts JM. Chapter 11. Defects in placentation and placental perfusion. Hypertension in Pregnancy. 2nd Edition. Ed: Lindheimer MD, Roberts JM, Cunningham FG. Stamford, Connecticut: Appleton & Lange; 1999: 377–394.

[47] Zhou Y, Damsky CH, Fisher SJ. Preeclampsia is associated with failure of human cytotrophoblasts to mimic a vascular adhesion phenotype. One cause of defective endovascular invasion in this syndrome? *J Clin Invest.* 1997;99:2152–2164.

[48] Kingdom JCP, Kaufmann P. Oxygen and placental villous development: Origins of fetal hypoxia. *Placenta.* 1997;18:613–621.

[49] Benirschke K, Kaufmann P, Baergen R. Chapter 15. Classification of villous maldevelopment. Pathology of the human placenta. 5th Edition. Ed: Benirschke K, Kaufmann P, Baergen R. 2006:491–518.

[50] Demir R, Seval Y, Huppertz B. Vasculogenesis and angiogenesis in the early human placenta. *Acta Histochem.* 2007;109(4):257–265.

[51] Demir R, Kayisli UA, Cayli S, Huppertz B. Sequential steps during vasculogenesis and angiogenesis in the very early human placenta. *Placenta.* 2006;27(6-7):535–539.

[52] Hanahan D. Signaling vascular morphogenesis and maintenance. *Science.* 1997;277 (5322): 48–50.

[53] Ribatti D, Vacca A, Nico B, Ria R, Dammacco F. Cross-talk between hematopoiesis and angiogenesis signaling pathways. *Curr Mol Med.* 2002;2(6):537–543.

[54] Olofsson B, Pajusola K, Kaipainen A, von Euler G, Joukov V, Saksela O, Orpana A, Pettersson RF, Alitalo K, Eriksson U. Vascular endothelial growth factor B, a novel growth factor for endothelial cells. *Proc Natl Acad Sci U S A.* 1996;93(6):2576–2581.

[55] Zhou Y, Bellingard V, Feng KT, McMaster M, Fisher SJ. Human cytotrophoblasts promote endothelial survival and vascular remodeling through secretion of Ang2, PlGF, and VEGF-C. *Dev Biol.* 2003;263(1):114–125.

[56] Vuorela P, Hatva E, Lymboussaki A, Kaipainen A, Joukov V, Persico MG, Alitalo K, Halmesmäki E. Expression of vascular endothelial growth factor and placenta growth factor in human placenta. *Biol Reprod.* 1997;56(2):489–494.

[57] Grimmond S, Lagercrantz J, Drinkwater C, Silins G, Townson S, Pollock P, Gotley D, Carson E, Rakar S, Nordenskjöld M and others. Cloning and characterization of a novel human gene related to vascular endothelial growth factor. *Genome Res.* 1996;6(2):124–131.

[58] Joukov V, Pajusola K, Kaipainen A, Chilov D, Lahtinen I, Kukk E, Saksela O, Kalkkinen N, Alitalo K. A novel vascular endothelial growth factor, VEGF-C, is a ligand for the Flt4 (VEGFR-3) and KDR (VEGFR-2) receptor tyrosine kinases. *EMBO J.* 1996;15(2): 290–298.

[59] Kalkunte SS, Mselle TF, Norris WE, Wira CR, Sentman CL, Sharma S. Vascular endothelial growth factor C facilitates immune tolerance and endovascular activity of human uterine NK cells at the maternal-fetal interface. *J Immunol.* 2009;182(7):4085–4092.

[60] Leung DW CG, Kuang WJ, Goeddel DV, Ferrara N. Vascular endothelial growth factor is a secreted angiogenic mitogen. *Science.* 1989;246(4935):1306–1309.

[61] Maglione D, Guerriero V, Viglietto G, Delli-Bovi P, Persico MG. Isolation of a human placenta cDNA coding for a protein related to the vascular permeability factor. *Proc Natl Acad Sci U S A.* 1991;88(20):9267–9271.

[62] Li H, Gu B, Zhang Y, Lewis DF, Wang Y. Hypoxia-induced increase in soluble Flt-1 production correlates with enhanced oxidative stress in trophoblast cells from the human placenta. *Placenta.* 2005;25(2–3):210–217.

[63] Munaut C, Lorquet S, Pequeux C, Blacher S, Berndt S, Frankenne F, Foidart JM. Hypoxia is responsible for soluble vascular endothelial growth factor receptor-1 (VEGFR-1) but not for soluble endoglin induction in villous trophoblast. *Hum Reprod.* 2008;23(6): 1407–1415.

[64] Achen MG, Jeltsch M, Kukk E, Mäkinen T, Vitali A, Wilks AF, Alitalo K, Stacker SA. Vascular endothelial growth factor D (VEGF-D) is a ligand for the tyrosine kinases VEGF receptor 2 (Flk1) and VEGF receptor 3 (Flt4). *Proc Natl Acad Sci U S A.* 1998;95(2): 548–553.

[65] Orlandini M, Marconcini L, Ferruzzi R, Oliviero S. Identification of a c-fos-induced gene that is related to the platelet-derived growth factor/vascular endothelial growth factor family. *Proc Natl Acad Sci U S A.* 1996;93(21):11675–11680.

[66] Baldwin ME, Halford MM, Roufail S, Williams RA, Hibbs ML, Grail D, Kubo H, Stacker SA, Achen MG. Vascular endothelial growth factor D is dispensable for development of the lymphatic system. *Mol Cell Biol.* 2005;25(6):2441–2449.

[67] Gu B, Alexander JS, Gu Y, Zhang Y, Lewis DF, Wang Y. Expression of lymphatic vascular endothelial hyaluronan receptor-1 (LYVE-1) in the human placenta. *Lymphat Res Biol.* 2006;4(1):11–17.

[68] Hauser S, Weich HA. A heparin-binding form of placenta growth factor (PlGF-2) is expressed in human umbilical vein endothelial cells and in placenta. *Growth Factors.* 1993;9(4):259–268.

[69] Ribatti D. The discovery of the placental growth factor and its role in angiogenesis: a historical review. *Angiogenesis.* 2008;11(3):215–221.

[70] Maglione D, Guerriero V, Viglietto G, Ferraro MG, Aprelikova O, Alitalo K, Del Vecchio S, Lei KJ, Chou JY, Persico MG. Two alternative mRNAs coding for the angiogenic factor, placenta growth factor (PlGF), are transcribed from a single gene of chromosome 14. *Oncogene.* 1993;8(4):925–931.

[71] Khaliq A, Li XF, Shams M, Sisi P, Acevedo CA, Whittle MJ, Weich H, Ahmed A. Localisation of placenta growth factor (PIGF) in human term placenta. *Growth Factors.* 1996;13(3-4):243–250.

[72] Achen MG, Gad JM, Stacker SA, Wilks AF. Placenta growth factor and vascular endothelial growth factor are co-expressed during early embryonic development. *Growth Factors.* 1997;15(1):69–80.

[73] Carmeliet P, Moons L, Luttun A, Vincenti V, Compernolle V, De Mol M, Wu Y, Bono F, Devy L, Beck H and others. Synergism between vascular endothelial growth factor and placental growth factor contributes to angiogenesis and plasma extravasation in pathological conditions. *Nat Med.* 2001;7(5):575–583.

[74] Gu Y, Lewis DF, Wang Y. Placental productions and expressions of soluble endoglin, soluble fms-like tyrosine kinase receptor-1, and placental growth factor in normal and preeclamptic pregnancies. *J Clin Endocrinol Metab.* 2008;93(1):260–266.

[75] Tsatsaris V, Goffin F, Munaut C, Brichant JF, Pignon MR, Noel A, Schaaps JP, Cabrol D, Frankenne F, Foidart JM. Overexpression of the soluble vascular endothelial growth factor receptor in preeclamptic patients: pathophysiological consequences. *J Clin Endocrinol Metab.* 2003;88(11):5555–5563.

[76] Maynard SE, Min JY, Merchan J, Lim KH, Li J, Mondal S, Libermann TA, Morgan JP, Sellke FW, Stillman IE and others. Excess placental soluble fms-like tyrosine kinase 1 (sFlt1) may contribute to endothelial dysfunction, hypertension, and proteinuria in preeclampsia. *J Clin Invest.* 2003;111(5):649–658.

[77] Levine RJ, Maynard SE, Qian C, Lim KH, England LJ, Yu KF, Schisterman EF, Thadhani R, Sachs BP, Epstein FH and others. Circulating angiogenic factors and the risk of preeclampsia. *New Engl J Med.* 2004;350(7):672–683.

[78] Levine RJ, Lam C, Qian C, Yu KF, Maynard SE, Sachs BP, Sibai BM, Epstein FH, Romero R, Thadhani R and others. Soluble endoglin and other circulating antiangiogenic factors in preeclampsia. *N Engl J Med.* 2006;355(10):992–1005.

[79] Armelin HA. Pituitary extracts and steroid hormones in the control of 3T3 cell growth. *Proc Natl Acad Sci U S A.* 1973;70(9):2702–2706.

[80] Gospodarowicz D, Cheng J, Lui GM, Baird A, Böhlent P. Isolation of brain fibroblast growth factor by heparin-Sepharose affinity chromatography: identity with pituitary fibroblast growth factor. *Proc Natl Acad Sci U S A.* 1984;81(22):6963–6967.

[81] Gospodarowicz D. Localisation of a fibroblast growth factor and its effect alone and with hydrocortisone on 3T3 cell growth. *Nature.* 1974;249(453):123–127.

[82] Olsen SK, Garbi M, Zampieri N, Eliseenkova AV, Ornitz DM, Goldfarb M, Mohammadi M. Fibroblast growth factor (FGF) homologous factors share structural but not functional homology with FGFs. *J Biol Chem.* 2003;278(36):34226–34236.

[83] Cao R, Bråkenhielm E, Pawliuk R, Wariaro D, Post MJ, Wahlberg E, Leboulch P, Cao Y. Angiogenic synergism, vascular stability and improvement of hind-limb ischemia by a combination of PDGF-BB and FGF-2. *Nat. Wed.* 2003;9(5):604–613.

[84] Gospodarowicz D, Cheng J, Lui GM, Fujii DK, Baird A, Böhlen P. Fibroblast growth factor in the human placenta. *Biochem Biophys Res Commun.* 1985;128(2):554–562.

[85] Shams M, Ahmed A. Localization of mRNA for basic fibroblast growth factor in human placenta. *Growth Factors.* 1994;11(2):105–111.

[86] Ferriani RA, Ahmed A, Sharkey A, Smith SK. Colocalization of acidic and basic fibroblast growth factor (FGF) in human placenta and the cellular effects of bFGF in trophoblast cell line JEG-3. *Growth Factors.* 1994;10(4):259–268.

[87] Antebya EY, Natanson-Yarona S, Hamania Y, Sciakia Y, Goldman-Wohla D, Greenfielda C, Arielb I, Yagela S. Fibroblast growth factor-10 and fibroblast growth factor receptors 1–4: expression and peptide localization in human decidua and placenta. *Eur J Obstet Gynecol Reprod Biol.* 2005; 119(1):27–35.

[88] Davis S, Aldrich TH, Jones PF, Acheson A, Compton DL, Jain V, Ryan TE, Bruno J, Radziejewski C, Maisonpierre PC and others. Isolation of angiopoietin-1, a ligand for the TIE2 receptor, by secretion-trap expression cloning. *Cell.* 1996;87(7):1161–1169.

[89] Maisonpierre PC, Suri C, Jones PF, Bartunkova S, Wiegand SJ, Radziejewski C, Compton D, McClain J, Aldrich TH, Papadopoulos N and others. Angiopoietin-2, a natural antagonist for Tie2 that disrupts in vivo angiogenesis. *Science.* 1997;277 (5322):55–60.

[90] Seval Y, Sati L, Celik-Ozenci C, Taskin O, Demir R. The distribution of angiopoietin-1, angiopoietin-2 and their receptors tie-1 and tie-2 in the very early human placenta. *Placenta.* 2008;29(9):809–815.

[91] Schiessl B, Innes BA, Bulmer JN, Otun HA, Chadwick TJ, Robson SC, Lash GE. Localization of angiogenic growth factors and their receptors in the human placental bed throughout normal human pregnancy. *Placenta.* 2009;30(1):79–87.

[92] Dunk C, Shams M, Nijjar S, Rhaman M, Qiu Y, Bussolati B, Ahmed A. Angiopoietin-1 and angiopoietin-2 activate trophoblast Tie-2 to promote growth and migration during placental development. *Am J Pathol.* 2000;156(6):2185–2199.

[93] Cascone I, Audero E, Giraudo E, Napione L, Maniero F, Philips MR, Collard JG, Serini G, Bussolino F. Tie-2-dependent activation of RhoA and Rac1 participates in endothelial cell motility triggered by angiopoietin-1. *Blood.* 2003;102(7):2482–2490.

[94] Kim KL, Shin IS, Kim JM, Choi JH, Byun J, Jeon ES, Suh W, Kim DK. Interaction between Tie receptors modulates angiogenic activity of angiopoietin2 in endothelial progenitor cells. *Cardiovasc Res.* 2006;72(3):394–402.

[95] Gale NW, Thurston G, Hackett SF, Renard R, Wang Q, McClain J, Martin C, Witte C, Witte MH, Jackson D and others. Angiopoietin-2 is required for postnatal angiogenesis and lymphatic patterning, and only the latter role is rescued by Angiopoietin-1. *Dev Cell.* 2002;3(3):411–423.

[96] Augustin HG, Koh GY, Thurston G, Alitalo K. Control of vascular morphogenesis and homeostasis through the angiopoietin-Tie system. *Nat Rev Mol Cell Biol.* 2009;10(3): 165–177.

[97] Suri C, Jones PF, Patan S, Bartunkova S, Maisonpierre PC, Davis S, Sato TN, Yancopoulos GD. Requisite role of angiopoietin-1, a ligand for the TIE2 receptor, during embryonic angiogenesis. *Cell.* 1996;87(7):1153–1155.

[98] Yancopoulos GD, Davis S, Gale NW, Rudge JS, Wiegand SJ, Holash J. Vascular-specific growth factors and blood vessel formation. *Nature.* 2000;407:242–248.

[99] Sato TN, Tozawa Y, Deutsch U, Wolburg-Buchholz K, Fujiwara Y, Gendron-Maguire M, Gridley T, Wolburg H, Risau W, Qin Y. Distinct roles of the receptor tyrosine kinases Tie-1 and Tie-2 in blood vessel formation. *Nature.* 1995;376:70–74.

[100] Geva E, Ginzinger DG, Moore DH 2nd, Ursell PC, Jaffe RB. In utero angiopoietin-2 gene delivery remodels placental blood vessel phenotype: a murine model for studying placental angiogenesis. *Mol Hum Reprod.* 2005;11(4):253–260.

[101] Carlson TR, Feng Y, Maisonpierre PC, Mrksich M, Morla AO. Direct cell adhesion to the angiopoietins mediated by integrins. *J Biol Chem.* 2001;276(28):26516–26525.

[102] Katz AB, Keswani SG, Habli M, Lim FY, Zoltick PW, Midrio P, Kozin ED, Herlyn M, Crombleholme TM. Placental gene transfer: transgene screening in mice for trophic effects on the placenta. *Am J Obstet Gynecol.* 2009;201(5):499.e1–8.

[103] Myatt L. Current topic: Control of vascular resistance in the human placenta. *Placenta.* 1992;13:329–341.

[104] Santos RA, Campagnole-Santos MJ, Andrade SP. Angiotensin-(1-7): an update. *Regul Pept.* 2000;91(1-3):45–62.

[105] Baker PN, Broughton Pipkin F, Symonds EM. Platelet angiotensin II binding and plasma renin concentration, plasma renin substrate and plasma angiotensin II in human pregnancy. *Clin Sci.* (Lond). 1990;79(4):403–408.

[106] Gant NF, Daley GL, Chand S, Whalley PJ, MacDonald PC. A study of angiotensin II pressor response throughout primigravid pregnancy. *J Clin Invest.* 1973;52:2682–2689.

[107] Gant NF, Chand S, Worley RJ, Whalley PJ, Crosby UD, MacDonald PC. A clinical test useful for predicting the development of acute hypertension in pregnancy. *Am J Obstet Gynecol.* 1974;120:1–7.

[108] Gant NF, Worley RJ, Everett RB, MacDonald PC. Control of vascular responsiveness during human pregnancy. *Kidney Int.* 1980;18(2):253–258.

[109] Brosnihan KB, Neves LA, Anton L, Joyner J, Valdes G, Merrill DC. Enhanced expression of Ang-(1-7) during pregnancy. *Braz J Med Biol Res.* 2004;37(8):1255–1262.

[110] Vickers C, Hales P, Kaushik V, Dick L, Gavin J, Tang J, Godbout K, Parsons T, Baronas E, Hsieh F and others. Hydrolysis of biological peptides by human angiotensin-converting enzyme-related carboxypeptidase. *J Biol Chem.* 2002;26(277):17.

[111] Wallukat G, Homuth V, Fischer T, Lindschau C, Horstkamp B, Jupner A, Baur E, Nissen E, Vetter K, Neichel D and others. Patients with preeclampsia develop agonistic autoantibodies against the angiotensin AT1 receptor. *J Clin Invest.* 1999;103:945–952.

[112] Dechend R, Homuth V, Wallukat G, Kreuzer J, Park JK, Theuer J, Juepner A, Gulba DC, Mackman N, Haller H and others. AT(1) receptor agonistic antibodies from preeclamptic patients cause vascular cells to express tissue factor. *Circulation.* 2000;101: 2382–2387.

[113] Dechend R, Viedt C, Muller DN, Ugele B, Brandes RP, Wallukat G, Park JK, Janke J, Barta P, Theuer J and others. AT1 receptor agonistic antibodies from preeclamptic patients stimulate NADPH oxidase. *Circulation.* 2003;107:1632–1639.

[114] Xia Y, Wen H, Bobst S, Day MC, Kellems RE. Maternal autoantibodies from preeclamptic patients activate angiotensin receptors on human trophoblast cells. *J Soc Gynecol Investig.* 2003;10:82–93.

[115] Hodari AA, Smeby R, Bumpus FM. A renin-like substance in the human placenta. *Obstet Gynecol.* 1967;29(3):313–317.

[116] Ihara Y, Taii S, Mori T. Expression of renin and angiotensinogen genes in the human placental tissues. *Endocrinol Jpn.* 1987;34(6):887–896.

[117] Kalenga MK, Thomas K, de Gasparo M, De Hertogh R. Determination of renin, angiotensin converting enzyme and angiotensin II levels in human placenta, chorion and amnion from women with pregnancy induced hypertension. *Clin Endocrinol* (Oxf). 1996;44(4): 429–433.

[118] Anton L, Merrill DC, Neves LA, Diz DI, Corthorn J, Valdes G, Stovall K, Gallagher PE, Moorefield C, Gruver C and others. The uterine placental bed Renin-Angiotensin system in normal and preeclamptic pregnancy. *Endocrinology.* 2009;150(9):4316–4325.

[119] Ito M, Itakura A, Ohno Y, Nomura M, Senga T, Nagasaka T, Mizutani S. Possible activation of the renin-angiotensin system in the feto-placental unit in preeclampsia. *J Clin Endocrinol Metab.* 2002;87(4):1871–1878.

[120] Herse F, Dechend R, Harsem NK, Wallukat G, Janke J, Qadri F, Hering L, Muller DN, Luft FC, Staff AC. Dysregulation of the circulating and tissue-based renin-angiotensin system in preeclampsia. *Hypertension.* 2007;49(3):604–611.

[121] Singh HJ, Rahman A, Larmie ET, Nila A. Raised prorenin and renin concentrations in preeclamptic placentae when measured after acid activation. *Placenta.* 2004;25(7):631–636.

[122] Anton L, Merrill DC, Neves LA, Stovall K, Gallagher PE, Diz DI, Moorefield C, Gruver C, Ferrario CM, Brosnihan KB. Activation of local chorionic villi angiotensin II levels but not angiotensin (1-7) in preeclampsia. *Hypertension.* 2008;51(4):1066–1072.

[123] Anton L, Brosnihan KB. Systemic and uteroplacental renin—angiotensin system in normal and pre-eclamptic pregnancies. *Ther Adv Cardiovasc Dis.* 2008;2 (5):349–362.

[124] Benoit C, Zavecz J, Wang Y. Vasoreactivity of chorionic plate arteries in response to vasoconstrictors produced by preeclamptic placentas. *Placenta.* 2007;28(5-6):498–504.

[125] Palmer MA, Piper PJ, Vane JR. Release of rabbit aorta contracting substance (RCS) and prostaglandins induced by chemical or mechanical stimulation of guinea-pig lungs. *Br J Pharmacol.* 1973;49(2):226–242.

[126] Yamamoto K, Ebina S, Nakanishi H, Nakahata N. Thromboxane A2 receptor-mediated signal transduction in rabbit aortic smooth muscle cells. *Gen Pharmacol.* 1995;26(7): 1489–1498.

[127] Nakahata N. Thromboxane A2: physiology/pathophysiology, cellular signal transduction and pharmacology. *Pharmacol Ther.* 2008;118(1):18–35.

[128] Hirata T, Ushikubi F, Kakizuka A, Okuma M, Narumiya S. Two thromboxane A2 receptor isoforms in human platelets. Opposite coupling to adenylyl cyclase with different sensitivity to Arg60 to Leu mutation. *J Clin Invest.* 1996;97(4):949–956.

[129] Offermanns S, Laugwitz KL, Spicher K, Schultz G. G proteins of the G12 family are activated via thromboxane A2 and thrombin receptors in human platelets. *Proc Natl Acad Sci U S A.* 1994;91(2):504–508.

[130] Shenker A, Goldsmith P, Unson CG, Spiegel AM. The G protein coupled to the thromboxane A2 receptor in human platelets is a member of the novel Gq family. *J Biol Chem.* 1991;266(14):9309–9313.

[131] Moncada S, Gryglewski RJ, Bunting S, Vane JR. A lipid peroxide inhibits the enzyme in blood vessel microsomes that generates from prostaglandin endoperoxides the substance (prostaglandin x) which prevents platelet aggregation. *Prostaglandins.* 1976;12: 715–737.

[132] Lafond J, Ayotte N, Brunette MG. Effect of (1-34) parathyroid hormone-related peptide on the composition and turnover of phospholipids in syncytiotrophoblast brush border and basal plasma membranes of human placenta. *Mol Cell Endocrinol.* 1993;92(2): 207–214.

[133] Lafond J, Moukdar F, Rioux A, Ech-Chadli H, Brissette L, Robidoux J, Masse A, Simoneau L. Implication of ATP and sodium in arachidonic acid incorporation by placental syncytiotrophoblast brush border and basal plasma membranes in the human. *Placenta.* 2000;21(7):661–669.

[134] Myatt L, Elder MG. Inhibition of platelet aggregation by a placental substance with prostacyclin-like activity. *Nature.* 1977;268:159–160.

[135] Walsh SW, Behr MJ, Allen NH. Placental prostacyclin production in normal and toxemic pregnancies. *Am J Obstet Gynecol.* 1985;151:110–115.

[136] Walsh SW. Preeclampsia: An imbalance in placental prostacyclin and thromboxane production. *Am J Obstet Gynecol.* 1985;152:335–340.

[137] Wang Y, Walsh SW, Kay HH. Placental lipid peroxides and thromboxane are increased and prostacyclin is decreased in women with preeclampsia. *Am J Obstet Gynecol.* 1992;167: 946–949.

[138] Bowen RS, Zhang Y, Gu Y, Lewis DF, Wang Y. Increased phospholipase A2 and thromboxane but not prostacyclin production by placental trophoblast cells from normal and preeclamptic pregnancies cultured under hypoxia condition. *Placenta.* 2005;26:402–409.

[139] Wang Y, Walsh SW, Guo J, Zhang J. The imbalance between thromboxane and prostacyclin in preeclampsia is associated with an imbalance between lipid peroxides and vitamin E in maternal blood. *Am J Obstet Gynecol.* 1991;165:1695–1700.

[140] Hauth JC, Goldenberg RL, Parker CR, Jr., Phillips JB, Copper RL, DuBard MB, Cutter GR. Low-dose aspirin therapy to prevent preeclampsia. *Am J Obstet Gynecol.* 1993;168: 1083–1093.

[141] Caritis S, Sibai B, Hauth J, Lindheimer MD, Klebanoff M, Thom E, VanDorsten P, Landon M, Paul R, Miodovnik M and others. Low-dose aspirin to prevent preeclampsia in women at high risk. *New Engl J Med.* 1998;338:701–705.

[142] Emeagi J, Patni S, Tikum HM, Mander AM. Low dose aspirin for preventing and treating pre-eclampsia. Author of editorial did not criticise studies' methodology. *BMJ.* 1999;319(7205):316.

[143] Kamio K, Sato T, Liu X, Sugiura H, Togo S, Kobayashi T, Kawasaki S, Wang X, Mao L, Ahn Y and others. Prostacyclin analogs stimulate VEGF production from human lung fibroblasts in culture. *Am J Physiol Lung Cell Mol Physiol.* 2008;294(6):L1226–1232.

[144] Spisni E, Griffoni C, Santi S, Riccio M, Marulli R, Bartolini G, Toni M, Ullrich V, Tomasi V. Colocalization prostacyclin (PGI2) synthase—caveolin-1 in endothelial cells and new roles for PGI2 in angiogenesis. *Exp Cell Res.* 2001;266(1):31–43.

[145] Sonveaux P, Martinive P, DeWever J, Batova Z, Daneau G, Pelat M, Ghisdal P, Grégoire V, Dessy C, Balligand JL and others. Caveolin-1 expression is critical for vascular endothelial growth factor-induced ischemic hindlimb collateralization and nitric oxide-mediated angiogenesis. *Circ Res.* 2004;95(2):154–161.

[146] Linton EA, Rodriguez-Linares B, Rashid-Doubell F, Ferguson DJ, Redman CW. Caveolae and caveolin-1 in human term villous trophoblast. *Placenta.* 2003;24(7):745–757.

[147] Zhao S, Gu Y, Lewis DF, Wang Y. Predominant basal directional release of thromboxane, but not prostacyclin, by placental trophoblasts from normal and preeclamptic pregnancies. *Placenta.* 2008;29 (1):81–88.

[148] Yanagisawa M, Kurihara H, Kimura S, Tomobe Y, Kobayashi M, Mitsui Y, Yazaki Y, Goto K, Masaki T. A novel potent vasoconstrictor peptide produced by vascular endothelial cells. *Nature.* 1988;332:411–415.

[149] Yanagisawa M, Masaki T. Molecular biology and biochemistry of the endothelins. *Trends Pharmacol Sci.* 1989;10(9):374–378.

[150] Barton M, Yanagisawa M. Endothelin: 20 years from discovery to therapy. *Can J Physiol Pharmacol.* 2008;86(8):485–498.

[151] Wilkes BM, Susin M, Mento PF. Localization of endothelin-1-like immunoreactivity in human placenta. *J Histochem Cytochem.* 1993;41(4):535–541.

[152] Malassiné A, Cronier L, Mondon F, Mignot TM, Ferré F. Localization and production of immunoreactive endothelin-1 in the trophoblast of human placenta. *Cell Tissue Res.* 1993;271(3):491–497.

[153] Barros JS, Bairos VA, Baptista MG, Fagulha JO. Immunocytochemical localization of endothelin-1 in human placenta from normal and preeclamptic pregnancies. *Hypertens Pregnancy.* 2001;20(1):125–137.

[154] Bilban M, Barth S, Cervar M, Mauschitz R, Schaur RJ, Zivkovic F, Desoye G. Differential regulation of endothelin secretion and endothelin receptor mRNA levels in JAR, JEG-3, and BeWo choriocarcinoma cell lines and in human trophoblasts, their nonmalignant counterpart. *Arch Biochem Biophys.* 2000;382(2):245–252.

[155] Cervar M, Kainer F, Desoye G. Pre-eclampsia and gestational age differently alter binding of endothelin-1 to placental and trophoblast membrane preparations. *Mol Cell Endocrinol.* 1995;110(1-2):65–71.

[156] Cervar M, Puerstner P, Kainer F, Desoye G. Endothelin-1 stimulates the proliferation and invasion of first trimester trophoblastic cells in vitro—a possible role in the etiology of pre-eclampsia? *J Investig Med.* 1996;44(8):447–453.

[157] Wilkes BM, Mento PF, Hollander AM, Maita ME, Sung S, Girardi EP. Endothelin receptors in human placenta: relationship to vascular resistance and thromboxane release. *Am J Physiol.* 1990;258:E864–E870.

[158] Faxén M, Nisell H, Kublickiene KR. Altered gene expression of endothelin-A and endothelin-B receptors, but not endothelin-1, in myometrium and placenta from pregnancies complicated by preeclampsia. *Arch Gynecol Obstet.* 2000;264(3):143–149.

[159] Faxén M, Nasiell J, Lunell NO, Blanck A. Differences in mRNA expression of endothelin-1, c-fos and c-jun in placentas from normal pregnancies and pregnancies complicated with preeclampsia and/or intrauterine growth retardation. *Gynecol Obstet Invest.* 1997;44(2):93–6.

[160] Cervar-Zivkovic M, Hu C, Barton A, Sadovsky Y, Desoye G, Lang U, Nelson DM. Endothelin-1 attenuates apoptosis in cultured trophoblasts from term human placentas. *Reprod Sci.* 2007;14(5):430–439.

[161] Buxton IL. Regulation of uterine function: a biochemical conundrum in the regulation of smooth muscle relaxation. *Mol Pharmacol.* 2004;65(5):1051–1059.

[162] Tichenor SD, Malmquist NA, Buxton IL. Dissociation of cGMP accumulation and relaxation in myometrial smooth muscle: effects of S-nitroso-N-acetylpenicillamine and 3-morpholinosyndonimine. *Cell Signal.* 2003;15(8):763–772.

[163] Luzi G, Caserta G, Iammarino G, Clerici G, Di Renzo GC. Nitric oxide donors in pregnancy: fetomaternal hemodynamic effects induced in mild pre-eclampsia and threatened preterm labor. *Ultrasound Obstet Gynecol.* 1999;14(2):101–109.

[164] Sabry S, Mondon F, Ferré F, Dinh-Xuan AT. In vitro contractile and relaxant responses of human resistance placental stem villi arteries of healthy parturients: role of endothelium. *Fundam Clin Pharmacol.* 1995;9(1):46–51.

[165] Leitch IM, Read MA, Boura AL, Walters WA. Effect of inhibition of nitric oxide synthase and guanylate cyclase on hydralazine-induced vasodilatation of the human fetal placental circulation. *Clin Exp Pharmacol Physiol.* 1994;21(8):615–622.

[166] Sprague B, Chesler NC, Magness RR. Shear stress regulation of nitric oxide production in uterine and placental artery endothelial cells: experimental studies and hemodynamic models of shear stresses on endothelial cells. *Int J Dev Biol.* 2010;54(2-3):331–339.

[167] Myatt L, Rosenfield RB, Eis AL, Brockman DE, Greer I, Lyall F. Nitrotyrosine residues in placenta: Evidence of peroxynitrite formation and action. *Hypertension.* 1996;28:488–493.

[168] Norris LA, Higgins JR, Darling MR, Walshe JJ, Bonnar J. Nitric oxide in the uteroplacental, fetoplacental, and peripheral circulations in preeclampsia. *Obstet Gynecol.* 1999;96(6): 958–963.

[169] Noris M, Todeschini M, Cassis P, Pasta F, Cappellini A, Bonazzola S, Macconi D, Maucci R, Porrati F, Benigni A and others. L-arginine depletion in preeclampsia orients nitric oxide synthase toward oxidant species. *Hypertension.* 2004;43(3):614–622.

[170] Ariel I, Hochberg A, Shochina M. Endothelial nitric oxide synthase immunoreactivity in early gestation and in trophoblastic disease. *J Clin Pathol.* 1998;51(6):427–431.

[171] Barut A, Harma M, Arikan I, Harma MI, Barut F. Endothelial nitric oxide synthase expression in gestational trophoblastic diseases. *Int J Gynecol Cancer.* 2010;20(3):337–340.

[172] Zimmermann K, Opitz N, Dedio J, Renne C, Muller-Esterl W, Oess S. NOSTRIN: a protein modulating nitric oxide release and subcellular distribution of endothelial nitric oxide synthase. *Proc Natl Acad Sci U S A.* 2002;99(26):17167–17172.

[173] Urata H, Kinoshita A, Misono KS, Bumpus FM, Husain A. Identification of a highly specific chymase as the major angiotensin II-forming enzyme in the human heart. *J Biol Chem.* 1990;265(36):22348–22357.

[174] Lüscher TF, Barton M. Endothelins and endothelin receptor antagonists: therapeutic considerations for a novel class of cardiovascular drugs. *Circulation.* 2000;102(19):2434–2440.

[175] Wang Y, Gu Y, Zhang Y, Lewis DF, Alexander JS, Granger DN. Increased chymotrypsin-like protease (chymase) expression and activity in placentas from women with preeclampsia. *Placenta.* 2007;28:263–269.

[176] Urata H, Kinoshita A, Perez DM, Misono KS, Bumpus FM, Graham RM, Husain A. Cloning of the gene and cDNA for human heart chymase. *J Biol Chem.* 1991;266:17173–17179.

[177] Caughey GH, Zerweck EH, Vanderslice P. Structure, chromosomal assignment, and deduced amino acid sequence of a human gene for mast cell chymase. *J Biol Chem.* 1991; 266(20):12956–12963.

[178] Benoit C, Gu Y, Zhang Y, Alexander JS, Wang Y. Contractility of placental vascular smooth muscle cells in response to stimuli produced by the placenta: roles of ACE vs. non-ACE and AT1 vs. AT2 in placental vessel cells. *Placenta.* 2008;29(6):503–509.

[179] Lee M, Sommerhoff CP, von Eckardstein A, Zettl F, Fritz H, Kovanen PT. Mast cell tryptase degrades HDL and blocks its function as an acceptor of cellular cholesterol. *Arterioscler Thromb Vasc Biol.* 2002;22(12):2086–2091.

[180] Lee M, Calabresi L, Chiesa G, Franceschini G, Kovanen PT. Mast cell chymase degrades apoE and apoA-II in apoA-I-knockout mouse plasma and reduces its ability to promote cellular cholesterol efflux. *Arterioscler Thromb Vasc Biol.* 2002;22(9):1475–1481.

[181] Mizutani H, Schechter N, Lazarus G, Black RA, Kupper TS. Rapid and specific conversion of precursor interleukin 1 beta (IL-1 beta) to an active IL-1 species by human mast cell chymase. *J Exp Med.* 1991;174(4):821–825.

[182] Pejler G, Rönnberg E, Waern I, Wernersson S. Mast cell proteases- multifaceted regulators of inflammatory disease. *Blood.* 2010;115(24):4981–4990.

[183] Wang Y, Zhang Y, Lewis DF, Gu Y, Li H, Granger DN, Alexander JS. Protease chymotrypsin mediates the endothelial expression of P- and E-selectin, but not ICAM and VCAM, induced by placental trophoblasts from preeclamptic pregnancies. *Placenta.* 2003;24: 851–861.

[184] Zhao S, Gu Y, Fan R, Groome LJ, Cooper D, Wang Y. Proteases and sFlt-1 release in the human placenta. *Placenta.* 2010; 31:512–518.

[185] Smith CH, Kamath SG. Trophoblast basal and microvillous membrane isolation. *Placenta.* 1994;15(7):779–781.

[186] Boyd CA, Chipperfield AR, Steele LW. Separation of the microvillous (maternal) from the basal (fetal) plasma membrane of human term placenta: methods and physiological significance of marker enzyme distribution. *J Dev Physiol.* 1979;1(5):361–377.

[187] Vanderpuye OA, Smith CH. Proteins of the apical and basal plasma membranes of the human placental syncytiotrophoblast: immunochemical and electrophoretic studies. *Placenta.* 1987;8(6):591–608.

[188] Sawicki G, Radomski MW, Winkler-Lowen B, Krzymien A, Guilbert LJ. Polarized release of matrix metalloproteinase-2 and -9 from cultured human placental syncytiotrophoblasts. *Biol Reprod.* 2000;63:1390–1395.

[189] Wang Y, Gu Y, Lewis DF. Endothelial angiotensin II generation induced by placenta-derived factors from preeclampsia. *Reprod Sci.* 2008;15(9):932–938.

[190] Wagner OF, Christ G, Wojta J, Vierhapper H, Parzer S, Nowotny PJ, Schneider B, Waldhäusl W, Binder BR. Polar secretion of endothelin-1 by cultured endothelial cells. *J Biol Chem.* 1992; 267(23): 16066–16068.

[191] Clozel M, Breu V, Burri K, Cassal JM, Fischli W, Gray GA, Hirth G, Löffler BM, Müller M, Neidhart W and others. Pathophysiological role of endothelin revealed by the first orally active endothelin receptor antagonist. *Nature.* 1993;365(6448):759–761.

[192] Red-Horse K, Rivera J, Schanz A, Zhou Y, Winn V, Kapidzic M, Maltepe E, Okazaki K, Kochman R, Vo KC and others. Cytotrophoblast induction of arterial apoptosis and lymphangiogenesis in an in vivo model of human placentation. *J Clin Invest.* 2006;116(10): 2643–2652.

[193] Red-Horse K. Lymphatic vessel dynamics in the uterine wall. *Placenta.* 2008;29(Suppl A): S55–59.

[194] Wetterwald A, Hoffstetter W, Cecchini MG, Lanske B, Wagner C, Fleisch H, Atkinson M. Characterization and cloning of the E11 antigen, a marker expressed by rat osteoblasts and osteocytes. *Bone.* 1996;18(2):125–132.

[195] Bailey D, Baumal R, Law J, Sheldon K, Kannampuzha P, Stratis M, Kahn H, Marks A. Production of a monoclonal antibody specific for seminomas and dysgerminomas. *Proc Natl Acad Sci U S A.* 1986;83(14):5291–5295.

[196] Kaufmann P, Stark J, Stegner HE. The villous stroma of the human placenta. Cell Tissue *Research.* 1977;177(1):105–121.

[197] Ordóñez NG. Podoplanin: a novel diagnostic immunohistochemical marker. *Adv Anat Pathol.* 2006(13):2:83–88.

[198] Alexander JS, Gu Y, Zhao S, Sun J, Groome LJ, Wang Y. Lymphatic endothelial marker D2-40 is expressed in the placenta villous stroma and its expression is reduced in placentas from women with preeclampsia. *Reprod Sci.* 2010;17(3):118A.

[199] Gluckman E, Broxmeyer HA, Auerbach AD, Friedman HS, Douglas GW, Devergie A, Esperou H, Thierry D, Socie G, Lehn P. Hematopoietic reconstitution in a patient with Fanconi's anemia by means of umbilical-cord blood from an HLA-identical sibling. *N Engl J Med.* 1989;321(17):1174–1178.

[200] Gahrton G, Björkstrand B. Progress in haematopoietic stem cell transplantation for multiple myeloma. *J Intern Med.* 2000;248(3):185–201.

[201] Lindvall O. Stem cells for cell therapy in Parkinson's disease. *Pharmacol Res.* 2003;47(4): 279–287.

[202] Goldman SA, Windrem MS. Cell replacement therapy in neurological disease. *Philos Trans R Soc Lond B Biol Sci.* 2006;361(1473):1463–1475.

[203] Matikainen T, Laine J. Placenta—an alternative source of stem cells. *Toxicol Appl Pharmacol.* 2005;207(2 (Suppl)):544–549.

[204] Portmann-Lanz CB, Schoeberlein A, Huber A, Sager R, Malek A, Holzgreve W, Surbek DV. Placental mesenchymal stem cells as potential autologous graft for pre- and perinatal neuroregeneration. *Am J Obstet Gynecol.* 2006;194: 664–673.

[205] Alviano F, Fossati V, Marchionni C, Arpinati M, Bonsi L, Franchina M, Lanzoni G, Cantoni S, Cavallini C, Bianchi F and others. Term Amniotic membrane is a high throughout source for multipotent Mesenchymal Stem Cells with the ability to differentiate into endothelial cells in vitro. *BMC Dev Biol.* 2007;21(7):11.

[206] Miki T, Lehmann T, Cai H, Stolz DB, Strom SC. Stem cell characteristics of amniotic epithelial cells. *Stem Cells.* 2005;23(10):1549–1559.

[207] Sakuragawa N, Kakinuma K, Kikuchi A, Okano H, Uchida S, Kamo I, Kobayashi M, Yokoyama Y. Human amnion mesenchyme cells express phenotypes of neuroglial progenitor cells. *J Neurosci Res.* 2004;78(2):208–214.

[208] Chen LB, Jiang XB, Yang L. Differentiation of rat marrow mesenchymal stem cells into pancreatic islet beta-cells. *World J Gastroenterol.* 2004;10(20):3016–3020.

[209] Ezquer FE, Ezquer ME, Parrau DB, Carpio D, Yañez AJ, Conget PA. Systemic administration of multipotent mesenchymal stromal cells reverts hyperglycemia and prevents nephropathy in type 1 diabetic mice. *Biol Blood Marrow Transplant.* 2008;14(6):631–640.

[210] Ryan JM, Barry FP, Murphy JM, Mahon BP. Mesenchymal stem cells avoid allogeneic rejection. *J Inflamm* (Lond). 2005;26(2):8.